The Many Futures of a Decision

ALSO AVAILABLE FROM BLOOMSBURY

Philosophy and Literature in Times of Crisis, Michael Mack
Simultaneity and Delay, Jay Lampert
The Ethics of Time, John Panteleimon Manoussakis
Enduring Time, Lisa Baraitser

The Many Futures of a Decision

JAY LAMPERT

BLOOMSBURY ACADEMIC
LONDON • NEW YORK • OXFORD • NEW DELHI • SYDNEY

BLOOMSBURY ACADEMIC
Bloomsbury Publishing Plc
50 Bedford Square, London, WC1B 3DP, UK

BLOOMSBURY, BLOOMSBURY ACADEMIC and the Diana logo are trademarks of Bloomsbury Publishing Plc

First published in Great Britain 2018

Copyright © Jay Lampert, 2018

Jay Lampert has asserted his right under the Copyright, Designs and Patents Act, 1988, to be identified as Author of this work.

For legal purposes the Acknowledgments on p. vi constitute an extension of this copyright page.

Cover image © Iulia Shevchenko / Alamy Stock Photo

All rights reserved. No part of this publication may be reproduced or transmitted in any form or by any means, electronic or mechanical, including photocopying, recording, or any information storage or retrieval system, without prior permission in writing from the publishers.

Bloomsbury Publishing Plc does not have any control over, or responsibility for, any third-party websites referred to or in this book. All internet addresses given in this book were correct at the time of going to press. The author and publisher regret any inconvenience caused if addresses have changed or sites have ceased to exist, but can accept no responsibility for any such changes.

A catalogue record for this book is available from the British Library.

A catalog record for this book is available from the Library of Congress.

ISBN: HB: 978-1-3500-4778-5
PB: 978-1-3500-4779-2
ePDF: 978-1-3500-4780-8
eBook: 978-1-3500-4781-5

Typeset by Newgen KnowledgeWorks Pvt. Ltd., Chennai, India

To find out more about our authors and books visit www.bloomsbury.com and sign up for our newsletters.

CONTENTS

Acknowledgments vi

Introduction: What is the temporal reference of a decision? 1
1. Sartre: Decisions and the unbound future 21
2. Husserl: Decisions and temporal overlap 53
3. Heidegger: The original decision to decide 85
4. Kierkegaard: Decisionism in religion. Infinite futures 105
5. Schmitt: Decisionism in politics. Sovereign moments. Habermas: Steering procedures and the term limits of a decision 125
6. Decision theory: Seriality effects on decision 155
7. Branching futures: Tense logic and multiple worlds 197
8. Hegel: Morality without decision. Derrida: Indecision Theory 225
9. Deleuze: Decision in the empty future. The virtual decision = X 251

Afterword 291

Notes 295
Bibliography 317
Index 329

ACKNOWLEDGMENTS

I would like to thank Duquesne University for generous research support for this project, including the President's Research Award in 2016. Jim Swindal (dean of arts), Ron Polansky (chair of philosophy), and all the faculty members and graduate students of the philosophy department have been excellent supporters and colleagues since I arrived at Duquesne in 2014. I would also like to thank the University of Guelph for generous sabbatical support in 2012–2013, when I was beginning this project. I thank my colleagues at the University of Guelph for many extremely helpful discussions of early versions of this project, particularly Karen Houle and John Russon, and also Jim Vernon at York University. I thank all my brilliant colleagues at Duquesne as well as the amazing graduate students at both Guelph and Duquesne for their contributions to my intellectual life.

I presented several pieces of this project at conferences over the last few years. I would like to thank Costas Boundas for organizing the Deleuze conference in Athens in 2015, where I presented part of Chapter 9; as well as Henry Somers-Hall for organizing a panel at SPEP in Atlanta in 2015, and the Lisbon Deleuze conference in 2013, where I presented other parts of Chapter 9. An early version of some of the Deleuze material in this book was published as "Deleuze's 'Power of Decision,' Kant's = X, and Husserl's Noema," in *At the Edges of Thought: Deleuze and Post-Kantian Thought*, edited by Craig Lundy and Daniela Voss. Edinburgh: Edinburgh University Press, 2015, 272–292. A different version of my SPEP paper on Deleuze will be published in *Deleuze Studies* in 2018. I also thank Jim Devin for organizing the Ontario-Quebec Hegel Organization meeting in Toronto in 2016, where I presented part of Chapter 8. I also thank Vedran Grahovac for organizing the Husserl Conference in Guelph in 2013, where I presented part of Chapter 2.

ACKNOWLEDGMENTS

I thank Jeff Mitscherling and Nicholas de Warren for disagreeing with me about Husserl; Jennifer Bates for disagreeing with me about Fichte and Hegel; Fred Evans for disagreeing with me about Derrida; and Dan Smith, Craig Lundy, Dan Selcer, and Sasa Stankovic for disagreeing with me about Deleuze.

The topics I am working on in this project are vast, and I take responsibility for all the pathways that I did not follow.

Thank you Hector.
Thank you Jennifer.

Introduction: What is the temporal reference of a decision?

When we make simple decisions, like the decision to wake up at 8:00 a.m. the next morning, we presuppose a simple linear model of the future. But when we make open-ended decisions, like the decision to get more involved in politics, which we might carry out in many different ways at many different times, we presuppose a more complex model of the future. We project a variety of possible futures. We can carry out the decision along many different pathways at once, which may converge or diverge at different points in time. Of course, not all possibilities for carrying out the decision can be articulated in advance, but this incomplete determinacy of the decision is generally understood at the time the decision is made, so its indeterminacy with regard to the future is part of the decision's determinacy.

There are certainly times when we make decisions without having foreseen all the consequences, and times when we feel bad about that upon later reflection. Sometimes we find ourselves doing something, and only retroactively attribute to ourselves an act of decision. Zizek assigns the "act of freedom" to "the time of the subject [that] is never 'present'—the subject never 'is', it only 'will have been': we never *are* free, it is only afterwards that we discover how we *have been* free."[1]

Zizek is overgeneralizing, but it is true that the moment of decision is a temporal floater, becoming explicit at various moments of action. Yet we know this fact about the ambiguous temporality of decision while we act now. We know in the present that we will be revising the present in the future. Failures of foresight are precisely implied by the fact that decisions, whenever they are made, are made about and for the future. After all, the mode of the future perfect, of decisions that will have been made as actions are carried out, captures more than retroactivity. It captures the way in which the retro is always already proto. It is not only that the present will be constituted as a past reference in the future, but also that the future is being constituted as a future reference in what is so far not past. The future may largely be about the past, but it is about the past while it is still future.

A decision projects a future that opens gaps in the time line. The idea of this study is that analyzing the wide range of temporal structures in decisions opens up a distinctive model of the future: A future composed of multiple, virtual time lines.

Most theories of decision concentrate on what the rational grounds for decision are, namely the evidence and the value assessments that build up into a rational decision. But my concern is exclusively with the nature of the future that multilayered decisions project, and this question of the future of a decision is surprisingly underrepresented in the literature on decision. Even when decision is treated as a problem of projected choice, that is, as *proairesis*, it is often treated less as a problem about the future, than as a problem about ends and means. And even when decision is treated under the thorny problem of predestination and divine foreknowledge, it is again often treated less as a problem about the future, than as a problem about free will. Decision only became a central topic in philosophy in the twentieth century. Before then, philosophy gave central roles to will and choice, to prediction and expectation, to imagination, desire, and hope. But decision has a different relation to the future than all of these. What can we learn about the future specifically from decision-making?

This book draws elements from some of the great twentieth-century philosophers. Most (but not all) of the figures I use come from the Continental tradition. I begin with a fascinating episode from Sartre's journals. When Sartre decided to go on a diet in 1942, he found himself unable to stick to his decision. His conclusion is that decisions never last into the future. Chapter 1 considers this deep challenge to my whole topic.

To get the topic moving, I use phenomenological approaches to decision and the future: A decision arises when a perceived actuality demands to be supplemented by one of many possibilities (Husserl); the decision kickstarts a subsequent history (Heidegger). I then consider several variations on "decisionism," the idea that decisions are spontaneous acts that cut an intention off from both the past and the future. I analyze the religious decisionism of Kierkegaard, as well as the political decisionism of Carl Schmitt. To bring the future back into decision, I draw ideas about the future from approaches to democracy in Habermas and Derrida. Although the issues of rational motivation in mathematical Decision Theory and Game Theory are outside of my concerns, I develop in my own way some important ideas in those disciplines, such as the distinction between serial and sequential decisions. In addition, I draw some ideas from tense logic to discuss how propositions about the branching future work. This will involve a brief discussion of possible worlds semantics and the "Many Worlds" interpretation of quantum physics. I use these interdisciplinary approaches to suggest a phenomenological picture of the branching, multiple, virtual world where decisions are to have their future. If decisions are really multiple and plastic, and endlessly revisable, then there is some overlap between multiple virtual decision-making and indecision. In the final chapter, I use several ideas in Deleuze to pull together the theory of the multiple future and the role of decision in constituting it.

My final conclusions will be: (a) Ethical decisions are lived out in virtual time more than in the "real world." (b) In the branching future, we can carry out any number of incompatible decisions.

The problem

The temporality of decision I am interested in is not the subjective sequence of decision-making (urge, deliberation, and decisive moment), but the temporal character of the decision object (which I call the temporal noema of the decision) that the decision makes a decision about: namely, the future alterations in the world that the decision aims to bring about. The tricky thing is that decisions are clearly about creating a new situation in the future, but they are also about the current situation to be changed. They are about determinate future states, but have to remain flexible enough to hold good as circumstances change, plans fail, or understanding improves. Of course, peoples' decisions are often made in concert or in competition, often simultaneously, that is, before each person knows the other's decision, as in the Prisoner's Dilemma, played over a series of rounds. The challenge of rationality is not just to make predictions in the face of uncertainty, but also to balance present and future values.

Decisions refer to what one intends to do in the future, but the specific moments in the future to which they refer are impossible to limit. Decisions are in that sense saturated with virtual futures. Their future referents are in one sense given simultaneously with the original decision, and in another sense, delayed. The openness to different possible applications of a decision, based on how the world happens to turn out, suggests that the time line in the decision's objective referent is set off from—parallel to, but not identical to—the actual future. The decision refers not just to a later moment of time, but to an autonomous temporal series, a virtual phase shift from actual time, yet which nevertheless has to join up again with actual time at those moments when the decision is at some point actually put into effect. This is the interesting twist on decisional time. If political decisions, for example, are open and delayed in this way, then politics has an alternative reality for its real object. Such a reality is not a utopian vision for the end of time, but works on the possible worlds that inhabit the alternate temporality coexisting in present intentionality.

And yet while a decision refers to a virtual world, a decision is not just a fantasy, or an idle promise to oneself, but a commitment. It sets a new sequence of acts in motion. But which acts, when? To go back to the simplest example, if I decide to wake up at 8:00 a.m. tomorrow, there is no puzzle about the temporal referent. If I decide in general to start waking up earlier for the indefinite future, there is an open set of referents—various morning times on various days—but still fairly closely circumscribed. But if I decide to do something about global warming, there is no specific action, and no well-marked range of actions, that decides success or failure. The decision is clearly void if I make no attempt at all to carry it out. But which future events are intended by the moment of decision? The moment of decision is paired with those later acts that, in retrospective fidelity, reenact it in practice. In Badiou's paradigm, for example, the nation-founding decision that declares that citizens are equal is grouped in a set with all of those moments in the future of that nation when further decisions have to be made about equality for illegal immigrants, or enemy combatants.

So on the one hand, decision aims to make the world different in the future; it aims at something different from the actual. Being decisive may even put the future more in doubt than it would have been if things had gone on as they were without any decision. On the other hand, the object of our decisions is the future of the real world. In the final analysis, decision is neither non-worldly nor worldly. It is about a quasi-actual world (in Husserl's sense of "quasi," developed in Chapter 2). Its objective contents are temporally variable, in-between imagined and actual time.

If I decide to do what I can about climate change, for example, I do not expect my first recycled bottle to cool down the world. I expect the world to continue pretty much as it would have otherwise, with a miniscule difference that may or may not contribute to larger divergence down the road. I look forward to a temporal succession in the real world that shifts just slightly act-by-act. Once I come to my decision's second enactment, and its hundredth (obviously decision phases are not countable in this simple way), there will have been uncountable unforeseeable divergences. And yet this differential is what the decision's forward-directed intentionality is about. The decision I make now is intended to affect an actual future, to treat actuality as a field of possibilities, and turn that possibility into a new actuality further in the future, which it expects to treat again as a possibility, and so on; but it does all this now, while possibility and actuality are both possible.

Decisions rely on possible world scenarios, and they are articulated with a tense logic that includes truth-value gaps, and yet they need real material futures as intentional referents to diverge from and to work with. I cannot decide to do something if the referent is too variable. I (probably) cannot decide to fall in love with whoever is the next person to get off the bus. But I can decide to settle down before I know whom to settle with. It is not quite that I never know what I am deciding to do at the moment when I decide. It is more that in decision I live through a series of ready possibilities that at each beat will be exactly what the actual world is at that time; but at that same time, each beat is chosen to phase out of what the actuality will be into what it should be. The time line of the decisional future is at each point out of phase with actual time by the length of just one act. If we take this idea of a virtual time line, and multiply it by all the decisions being made at a given present, and by all the variations on each decision being contemplated and enacted at the same time for the same future, then we will have a theory of the multiple virtual future.

The method

We can distinguish three philosophical approaches to the temporality of decision, focusing on (a) the past of a decision, namely the reasons for making a decision (including its historical frameworks, influences and motives, and rational criteria); (b) the present moment of a decision (whether the decision breaks the continuity of events or not); and (c) the future of a decision. All three are good topics. In this book, I restrict myself to the third, though this will mean that I will not focus on many otherwise important theorists of decision, from von Neumann to Badiou.

Three methodologies are employed in this book. (a) To gather together and analyze diverse phenomena of decision; (b) to explore discussions of decision and time in the history of philosophy, with emphasis on Continental philosophy since 1900; and (c) to synthesize the results into an overall philosophy of the future.

1. Different areas of decision-making have different patterns of future content, so I raise phenomena from both existential and economic decisions, political, judicial, and lifestyle decisions, the religious will to believe and the self-interested decisions of the Prisoner's Dilemma, oaths and promises, decisions with and without finite end points, decisions with diverging and converging sequences, decisions with overlapping options, as well as the power of indecision. The book contains concrete applications to such phenomena as democratic term limits, dieting, risk management, and time travel science fiction.

2. Decision is a central concept in each of the major schools of twentieth-century Continental philosophy. In Husserl, decision is the first step in the genesis of rationality. In

Heidegger, it is the founding act of being-in-the-world. In Sartre, decision represents the moment of anxiety. In Habermas, it is the steering apparatus for pluralistic social consciousness. In Derrida, decision cuts into signifiers and events in such a way as to render the decision itself undecidable. In Deleuze, it multiplies and serializes events. If decision for Husserl presents many options and selects just one, and decision for Deleuze takes one situation and fragments it into overlapping series, then the issue of decision weaves a typical line through the history of twentieth-century thought.

In part, what explains the lack of attention to the decisional future is that philosophers interested in decision have focused more on the rationality of decisions than on their temporality. Decision Theorists have focused on problems of evidence and goal assessment—on what a person knows while making the decision combined with what she values as an end result—rather than on how the decision-maker thinks the decision will carry out changes in the world step by step in the future. My approach is to focus on the future in the decision, independent of the rational evidence the decision-maker does or does not possess. If we ask what a decision's content is, what a decision is about, then at least in one sense, a decision is about time, and its content is the future.

3   My plan is thus to focus exclusively on the futural projection of decision and its distinctive complex timescape.

Other future-oriented noeses

The premise of this book is that the way decisions refer to the future is different from the way other noeses and propositional attitudes refer to the future. Predictions, desires, oaths, hopes, plans, deals, operational codes, structurational processes, and many other forms of individual consciousness and institutional intersubjectivity also refer to the future, but with either more or less determinacy than decisions do. Each attitude captures some authentic relation to the future, so I do not absolutely privilege any one of them for having a correct vision of the future. If it turns out that there are different models of the future that emerge out of different noetic attitudes, with no single overarching model of the future, then that itself will be an interesting result. The "future" might well turn out to be an equivocal term that covers many different temporal structures. Still, decision is a promising comportment toward the future. Decision has enough openness to express the fact that the future has not happened, and yet not so much openness as to prevent us from aiming at the future. In that way, decision captures the distinctive subtlety of future time.

Decisions and plans: In making a plan, we set out a sequence of acts, though we know it can go wrong; in making a decision, we commit ourselves to revising the plan to get to the desired end. A flexible plan may prepare a flowchart of options for as many chance events as can be preimagined; whereas a decision has to hold even if there are unpredicted nodes of chance, even if we do not know what kinds of states the chance nodes will determine, or how many there will be, or where they will figure in the flow. Generally, we want our decisions to hold in spite of fairly major chance interruptions. In the extreme (but common) case, a decision also anticipates chance events that will make us decide to cut the previous decision short. Decision, in short, has to allow for meta-chance, not just for controlled chance.

Decisions and utopias: Utopian descriptions sometimes describe alternate realities that supposedly exist somewhere (*Erewhon* somewhere), or ideal possibilities that might come about (the *Republic*), or ideals that supposedly will necessarily come about (scientific socialism). Utopias are temporally ambiguous. They are often narrated more as a place than a time, displacing a dream of future time onto a distant society on or off planet. But the time–space displacement can be deployed in various ways. When the utopian place is fictional, and unreachable, sometimes the idea is that the utopia already exists in present time, though elsewhere. But sometimes, describing a fictional utopia suggests that the best of the world is infinitely deferred in time. In either case, utopias only indirectly call for decision. Plato calls for a judgment about the Republic, not for a decision. Marx and Engels' manifesto calls for engagement, and engagement is closer to decision than judgment is, but is still not quite the same thing. Manifestos are generally more about justification or command than decision, and more about joining a current movement than about preparing for future contingents. Still, a communist, an environmentalist, or an aesthetic manifesto might fit the decision mode.

Decisions and intentions (not speaking here of phenomenological intentionality, but of intending to do something): For G. E. M. Anscombe,[2] analyzing intention means finding the sense of the question "Why?" Her goal is to "distinguish actions which are intentional from those which are not" (section 5). Largely, this means distinguishing causes from reasons. For her, intention is not primarily about the fact that "people give accounts of future events in which they are some sorts of agents" (section 3). I think it is fair to say that having an intention does not always require giving an account of future events; but making a decision does require this, which is why I am interested in decision.

Some of Anscombe's analysis of intention is nevertheless relevant to decisional futures. For example, to adapt a point she makes about intention, a single decision can sometimes be described in two ways, which then have different temporal connotations. The "decision to go to SPEP in 2015" can be restated as the "decision to drive fourteen hours to Atlanta." Although they have the same reference and the same truth-value, the two expressions have different futural senses. As Anscombe says, I might recognize that I am making one of those decisions but not notice that I am also making the other. I might not, then, realize, or think about, the futural reference in my decision.

Of course, not all analytic philosophers agree with Anscombe's hard distinction between intent and future reference. Andy Hamilton, for example, proposes a "future-outer thesis."[3] On this thesis, intentions in decision-making are not about inner states of mind existing in the present; they are predictions about factual states in the future (23). An intention is not a prediction of facts about the future (he agrees with Anscombe up to this point), but it is a prediction about how one intends to act in the future, and therefore, the intention is a kind of "knowledge" of at least one thing that will happen in the future. What Hamilton calls the "Decision Principle" sometimes attributes too much certainty and invariability to decisional futures. For example, he says that, "having decided the question whether I will A/am going to A, there is no further question for me to decide, concerning whether I intend to do A" (25). Hamilton also thinks that intentions cannot be unconscious, and that if I have an intention and act on it, I cannot be surprised that I am doing that action. This thesis too might be too strong; in fact, the possibility of being surprised at oneself would itself be an interesting subtopic in a theory of the future of decisions.

I mention a few of these tricky cross-noesis problems here not because they will all be solved in this book, but to suggest how I will be looking for similar micro-problems in the course of exploring the decisional future.

Preview of chapters

Chapter 1. Sartre: Decisions and the unbound future

Argument: A decision is an act of will directed at the future, but since a decision cannot control what the will will do in the future, it is only "about" the future in a defective way.

Sartre's *Being and Nothingness* famously describes a gambler who decides to quit, but who finds by the very next temptation that he has to make the decision all over again. In his *War Diaries*, Sartre analyzes his own endlessly problematic decision to go on a diet. I compare Sartre with William James' psychology of the flow of decision-making.

If a decision has to be remade at every moment, in what sense is it a decision at all? If there is no future to the act of commitment, then what is the temporal content of the decision? If the future has no subjective future, how can it objectively be about the future? One might conclude that decisions are simply reflections of a current state of mind, and have no intentional relation to the future at all. Some of Sartre's theses no doubt go too far, but what we learn from Sartre is that decisions have an odd sort of future, a future that is in one sense discontinuous with the present.

Chapter 2. Husserl: Decisions and temporal overlap

Argument: The phenomenology of overlapping experiences shows how decisions can project into many possible future time lines at once.

To put the question in Husserlian terms: what is the futural noema of a decision?

Husserl's *Experience and Judgment* analyzes decisions at two stages. On the one hand, decision strives to transform the uncertainties of the life-world into fixed and singular judgments. On the other hand, once they are made, even the most fixed decisions lose their certainty over time as new experiences support different alternatives. Consequently, decisions are always being juxtaposed with overlapping decisions and counter-decisions. Therefore, in the final analysis, singular judgments project into multiple future time lines.

The theme of overlapping experiences is a recurring motif in *Experience and Judgement*. Structures of overlap (*überschiessen*), juxtaposition (*nebeneinandersetzen*), and superimposition provide categories for understanding how decisions project into different overlapping futures—real futures with different contents and sequences posited over the same time field. The life-world of a decision is a real but virtual situation, taking place in overlapping times.

Chapter 3. Heidegger: The original decision to decide

Argument: Our sense of time and history originates with a decision. Heidegger's strong sense of the history of decision leaves him with a weak sense of the future of decision.

Heidegger is not very interested in everyday decisions and the future actions they anticipate and commit to. He is interested in the originary "de-cision," the ontological condition of everyday decision, the source of authenticity and of all engagement in-the-world, all of our sense of history and care, as well as the foundation of our sense of the future. The de-cision in question is not any particular decision, but is not a universal or generic decision either: it is the decision to decide. On the one hand, this is as future-directed as it is possible to be, since what *Being and Time* calls "resoluteness" kicks off the very possibility of all future decisions. On the other hand, the original de-cision in Heidegger's *Beiträge* is so originary that it leaves little room for specifying which actual futures are faithful to the original commitment, or for specifying the structure of future temporality.

The idea that decision has value independent of whatever future(s) it brings about, can be labeled "decisionism," and Heidegger has been accused of this. The charge is not quite right in his case, as Heidegger has many constraints on what can count as the originary de-cision. Still, it is true that Heidegger's account of decision has little to say about the future of enactments, assessments and revisions, reversals and accelerations, divergences and convergences, second guesses, and rushes to judgment.

Chapter 4. Kierkegaard: Decisionism in religion. Infinite futures

Argument: Decisions of faith behave as though they are binding for eternity, but the hermeneutics of retroactive interpretation multiplies the levels of the decision's future.

In *Concluding Unscientific Postscript*, Kierkegaard declares that, "the leap is the category of decision," the "eternal decision" (99). But because it is "postponed," it is also a "decision within time." I compare the future in Kierkegaardian decision with the future in Pascal's wager, the future in Kant's moral postulate of hope, and the future in an Ancient Egyptian moral text on eschatology.

A decision, says Kierkegaard, must not be concerned with "the slightest detail" of any concrete "case in point." Such decisions, unconcerned with world history, sound decisionistic. But Kierkegaard's association of decisions with promises suggests something else. Perhaps decisions are not simply indifferent to future outcomes, but pertain to a different kind of future.

A clue may be found in the Preface of the *Postscript*, where Kierkegaard (under pseudonym) recounts the story of writing the book. He had once promised a second volume, which he never wrote. And a reviewer had promised to review the first volume, but never did. Some years later, many readers remember that once-promised review, as if it had been written, and also have the false memory of having read Kierkegaard's never written second volume. Kierkegaard's story is ironic, but it serves as an analogue for the relations between decision, promise, and faith. A decision has a future, but its future is mapped not by its factual effects, but by the subjectivity of the readers who make their own decisions about where to find its sequels.

A decision's futural content thus depends on the content of future decisions. In a comparable way, Robert Sokolowski's religious phenomenology hangs on recursive, responsive, and interactive decision-making, speculating that recurring acts of decision bind the eternal and the temporal human together.

Chapter 5. Schmitt: Decisionism in politics. Sovereign moments. Habermas: Steering procedures and the term limits of a decision

Argument (a): Political decisionism is even more atemporal than religious decisionism, but there is no future in it; it is easy to refute.

Carl Schmitt's *Political Theology* claims that any genuine decision makes an exception to the rules and traditions of normal social life. Pure decisions are sovereign. This chapter explores the temporal implications of Schmitt's political ideal: not only the political disengagement from history, a consequence that is much discussed, and might not be entirely a bad idea; but also the political disengagement from the future. Schmitt's ideal decision is so ahistorical and counter-actualizing, that there is no legitimate follow-up, and no commitment to any future layout, only a series of equally authoritarian decrees in stand-alone present moments. That might be what Schmitt intended. But in the end, abstracting from futural implications removes all meaning whatsoever from a decision. In any case, Schmitt's empirical evidence for thinking that purely sovereign decisions have been made in the past, and were being made in Germany at the time of his writing (1934), is not convincing. There never were, and never could be, the future-free decisions that Schmitt thematizes.

Argument (b): In political decisions, the degree of future influence that a decision has is itself a matter for decision.

Habermas's *Legitimation Crisis* criticizes Schmitt and Luhmann's decisionism for failing to include the social processes that steer a plurality of interests into a decision with practical effects.

In fact, Niklas Luhmann's *Risk: A Sociological Introduction* contains something important for decisional futures: that risk is an irreducible and desirable feature of social decision-making. "Risk" adds an important element of decisional futuricity, distinguished from both uncertainty and revisability. Risk is a feature of the strong temporality of decisions, rather than of the weak rationality of decisions.

Habermas analyzes social systems of communication and controlled dispute, and of constructive norms. I develop implications of these steerage procedures for decisional futures. Drawing points from Derrida, as well as from applied political theory, I apply the "until future notice" feature of democratic decision procedures, to topics from term limits to judicial decision-making.

Chapter 6. Decision theory: Seriality effects on decision

Argument: Game-theoretical analyses of sequential and iterated decisions structure the branching future of decisions.

Decision Theory, or Game Theory, is largely concerned with maximizing the rationality of decisions. That is not the concern of this book. Nevertheless, Decision Theory introduces many subtopics that I will use to study the futural structure of decision contents.

For example, Decision Theory distinguishes a sequential decision (different subdecisions adding up to one whole decision) from an iterated decision (the same complete decision repeated on many occasions). Different patterns of multi-step decisions allow for different ways of revising expected and desired futures over time. In general, each decision envisages not just one desired future, but a many-leveled sequence of revisable futures within the selected future, or many selections of the future under changing conditions. Most of the actual future choices will not be articulated in advance, since they depend on evidence, ideas, and values that will only arise at later times. Some Decision Theorists hold that reasonable decision-making under uncertainty ought to show a preference for options that are easily reversible with changing conditions. This can be compared to Rawls' idea of "saving for the future," as a mode of social justice.

One important distinction is between decisions whose number of steps is finite and those that will be indefinitely reiterated. Prisoners' Dilemma decisions give suboptimal results when each participant knows there will only be one chance to make their decision (i.e., one chance to trust or betray each other). When there are indefinitely many occasions for changing one's mind, participants can experiment with mutual trust without the risk of immediate failure. Drawn-out, delayed, and indefinitely futural definitions thus give more chances for cooperative decision-making. This is in part because the value of having a good "reputation" can play a role in decision-making only over time, and in part because over time one can afford to ignore one's opponent's temporary bad choice, or even to allow for one's opponent to have made a purely accidental mistake and forget about it. One might have thought that "Backward Induction," that is, starting with a defined future and reading back to what one should do now, which Aristotle calls "deliberation," should yield thoughtful, future-directed decision-making. But in practice, indefinitely drawn-out decision procedures, aiming at a future so futural that there is no defined future state to induce backward from, yields decisions with more future content, as well as more chance of success.

There are many interesting phenomena in Decision Theory from which I draw further elements of decisional futuricity. One is "Newcomb's Problem," which involves how to take into account evidence that looks like foreknowledge of the future. Another problem arises in multistep decisions where the decision-maker needs to know which step she has arrived at, but where different steps are so similar that there is no way to be sure how far along in the history of the decision she has traveled and how far she has yet to travel. Other problems involve applications of Decision Theory to economics, applications of Systems Theory to urban development, and applications of Artificial Intelligence to go-playing computers.

Chapter 7. Branching futures: Tense logic and multiple worlds

Argument: The structure of branching futures is a condition for a theory of the future of decisions.

Because a decision has many futures, and yet it is not decided at the time of decision which of these will be enacted in which way, a tense logic for future contingents should help to describe it. I use Richmond Thomason's "Indeterminist Time and Truth-Value Gaps" (1970) as a basis. I take a cue from a paper by Briggs and Forbes (2012), and compare two models of the branching future: C.

D. Broad's (1923) "Growing Block" model of the future, and Storrs McCall's "Shrinking Tree" (1994) model. I adapt Dummett's essay "The Reality of the Past" (1978) into a discussion of the reality of the future. I defend "realism" with respect to future branches, to go along with Lewis' "modal realism" for possible worlds.

I also make use of other approaches to branching futures: Belknap and Perloff's (1988) subtle version of agency; David Deutsch's (1997) "Many Worlds" interpretation of quantum physics; as well as science fiction speculation around time travel into the future.

Chapter 8. Hegel: Morality without decision. Derrida: Indecision Theory

Argument: If decision leads in different directions at the same time, one might wonder whether (a) decision does not really play a useful role in action after all; or (b) whether all attempts to make decisions tend to collapse into (and maybe ought to be protected from) indecision and inaction.

Argument (a): Few philosophies of moral action before the twentieth century focus on decision. I have chosen Hegel's chapter on "Morality" from his *Phenomenology of Spirit* as a sample case. When Hegel does occasionally mention decision, he has mostly disparaging things to say about its role in morality, often precisely because decision-making gets caught up in alternatives that confound the unity of the good will. I try to show how a dialectical account of decision would use this defect, or challenge, to decision-making to pluralize morality in general.

Argument (b): Decisions retain alternative branching futures even while they actualize just one of them. Therefore, decision is a special case of indecision. Morally and politically, indecision might generally (but not always) be preferable to decision.

In the "Afterward" to *Limited Inc*, Derrida makes a crucial distinction between the undecidable and the indeterminate. Critics of Derrida tend to miss this distinction, but it is the key to seeing that deconstruction is not a kind of relativism based on indeterminate vagueness in meaning, but an analysis of multiplicity and deferral in communication, intentionality, and context.

I ask whether Derrida sides with a kind of decisionism closer to Schmitt (about whom Derrida has a mixed evaluation in *The*

Politics of Friendship), or a kind of unfinished democratic theory closer to Habermas. Derrida's conception of "Messianism without a Messiah" in *Spectres of Marx* can sound like the former. But his analysis of telecommunication in "Signature Event Context" adds up to something more like the latter. The difficulty of deciding whether deconstruction is more like decisionism or more like hermeneutics, is due to the overlap of decision and indecision.

In one sense, there is more futural content, and sometimes more determinacy, in indecision than in decision. People often say, either to criticize or to moderate Derrida, that eventually a person has to make a decision. Why? There are good reasons, suggested by some current Italian political philosophers, to imagine that a life of indecision, and a democracy where decisions are delayed, postponed, and kept revisable to a maximal degree, would provide the best life. Morality might indeed require rigorous indecision. A well-designed Decision Theory might indeed favor indecision as a way of maximizing value. Alternatively, an Indecision Theory might have its own way of analyzing and pursuing value. It might be best to say that a morality of futuricity hangs on the undecidable relation between decision and indecision. That would mean that one could commit oneself decisively to future action without excluding alternatives. This should bring together Hegel's attempt to exclude decision from moral action, and Derrida's deconstruction of both decision and indecision.

Chapter 9. Deleuze: Decision in the empty future. The virtual decision = X

Argument: Decisions are turning points from which event-lines, each extending in a certain direction, are distributed over the same duration in many directions and at different tempos.

When Deleuze discusses "the power of decisions" in *Difference and Repetition* and *Logic of Sense*, he describes decisions as singularities emitting diverging lines into the future. Decision is sharply distinguished from "judgment," which cuts off deliberation and imposes a limit. Deleuze says the "future is the proper place of decision." In *The Fold*, Deleuze uses Leibniz to describe Adam's decision whether or not to sin as a "pure game" that diverges into different possible and incompossible worlds.

A turning point is carried by relays, shaped by direction and vicinity, density and distance, and time to arrival at destinations by different routes. Like any object that remains open to many future attributions, which Kant calls an object = X, a decision opens onto many future postscripts and re-orientations. I will call the object of such an act: a decision = X.

Deleuze's account of decisional multiplicity is complicated by his analysis of the future as "the pure and empty form of time" in *Difference and Repetition*. If the future is really empty, then there is no intentional possibility for future-directed decisions. But Deleuze's account of empty form can be read as a region of virtuality, and in that way, can add to rather than subtract from the layered future.

Because the meaning of a decision includes many possible directions, a possible worlds theory should help to explain its form. I compare David Lewis' *On the Plurality of Worlds* (1986) with Deleuze's theory of incompossible worlds. Temporal sequences are not always at the forefront of possible worlds theories, but they should be. Lewis argues that it is absurd to say that a person can live in many possible worlds at once, but I construct a Deleuzian argument to the effect that that is the norm.

The question of whether a decision is therefore one or many, is complicated: by moving into the future through a turning point, does one remain in the same world, or bifurcate into many possible worlds, or bifurcate through virtual time lines within a single actual world? I argue for the latter.

In each chapter, I get into the details of the philosopher(s) treated there. Although I have chosen texts that pertain to decision, not every subtopic in those texts is equally relevant to the topic of decision. But after each textual study, I promise to return to the main path. That degree of partial convergence is about what we should expect from a multileveled branching series, namely diverging and converging nodes as the futures of decisive turning points, with subdecisions and virtual futures layered by backtracks and reboots.

Conclusion

A decision is structured by its future reference, and its meaning is distributed among a plurality of virtual time lines. There are many patterns of time line convergence and divergence, and different decisions have different types of future reference. But all these patterns can be schematized under forms of virtual temporality. Decisions are made for the world in its virtual time, and the world in virtual time is the only world in which we can exercise moral responsibility.

CHAPTER ONE

Sartre: Decisions and the unbound future

Sartre's thesis is that a decision to act in a certain way in the future cannot determine one's actions in the future. There may be a sense in which this is true. However, in its extreme form, it would mean that decisions have no future. If a decision has no future, and pertains only to the moment it is made, it breaks the before–after relation that would seem essential to decision-making. On the other hand, if the decision only determines the future, but does not express a futural orientation in the present, it again breaks the before–after relation. Both these ways of describing time and decision remove time from the decision. Such a decision might hold some appeal as a radical break, but it would not be radically temporal. I am looking for the futural time inside the decision content.

On December 17, 1939, Jean-Paul Sartre decided to go on a diet. France and Germany had gone to war in September. Sartre had done his military service back in 1929 in the meteorological corps, so he was remobilized to do the same job in an Artillery Headquarters in the town of Morsbronn on the border. Unexpectedly, the Germans did not invade for many months. This period is known as the *drôle de guerre*, "The Phoney War." As he was neither a regular soldier nor an officer, and his job was easy and unimportant, he had a lot of time. He kept a journal, translated as the *War Diaries*.[1] These contain drafts for *Being and Nothingness*, reflections on literature and politics, and phenomenological gossip about his troop. He assesses his childhood and love life with the self-critical yet self-aggrandizing penetration that made him not only the inventor of bad faith but also the object of its most honest and ironic application. He rigorously challenges his own honesty in relation to the future. He indulges in illusions of decision futurity, but then he analyzes his indulgences, almost as if he indulged just to critique himself, then critiques his indulgence in the pleasure of self-critique.

Sartre wrote in his barracks and in the local cafés, where he ate all too well.[2] He decided to go on a diet.

I will discuss Sartre's analysis of decision, using the five page diet entry in the *War Diaries*. First, I will discuss earlier entries on decision and time. Then I organize Sartre's description of the diet into two steps. I call step one of the dieter's decision problem: Breakfast and Time and step two: Lunch and the Other. There is no Dinner. I end with Sartre's image that a decision is like arranging a rendezvous with one's future self.

A decision obviously expresses an intention toward the future. However, Sartre finds, theoretically and practically, that it cannot hold in its future. *Being and Nothingness*[3] treats similarly the gambler whose decision to quit, even if carried out with total success for a day, might as well never have been made by the first moment of temptation the next morning. There are a number of differences between the gambler and the dieter. Sartre's gambler is an idealized object lesson that makes a definite claim that decisions have no future; the dieter is Sartre's actual experience, which in a less definite way, rides the waves of decision and indecision. The gambler has "anguish in the face of the past;" the dieter's anguish is for the future. Perhaps neither the gambler nor the dieter is a general paradigm of real-world decision, after all as they are both abstainers. A revolutionary's decision would be a different paradigm: proactive, although still not easily enacted. The decision to write a journal, which Sartre also anguishes about, is yet another paradigm: more easily enacted, but also more "humble" (Sartre 68). Still, all such personae—the gambler, the dieter, and the writer—have in common that they find their decisions melting away. The past melts once the future becomes present. In other words, the past becomes past.

In this and the subsequent chapters, I am heading toward three counterintuitive theses. First, we do *not* have to come eventually to a decision and stick with it. The idea that we do is commonly heard in politics. However, term limits demonstrate that political decisions cannot reasonably last very long. Maybe indecision, even daily revision, is not bad for politics. In addition, even if some decisions ought to stick, they still require further decisions. Sokolowski, for example, says that, "serious decisions"—not like the decision to buy a car, but like "a decision to marry a certain person, the decision to adopt this child, or the choice to enter a particular religious community" (Sokolowski 76)[4]—"involve further decisions and choices to be made in response." Decision A is serious if it includes both itself and Decision B. Or in other words, a serious decision is a decision made at two separated moments of time.

My second thesis is that when we do make a decision, it is not about the real world. Our moral responsibility is to make decisions about the virtual world. It sounds strange to leave the actual world in order to affect it, to be moral by running away, but gradually I would like to show this.

The third thesis is that virtual time contains a plurality of time lines, that the future branches not just to different events that might occupy the same time line, but that decision takes us into multiple futures at once. I think that Sartre's theory that decisions cannot operate on the future is distorted just because he does not sufficiently pluralize the temporal content of decisions. Even so, Sartre's premise that the dieter's decision fades is eminently plausible, and his text contains all sorts of great ideas, both large and small.

When defining decision, Sartre occasionally makes use of Husserl's vocabulary of the noema. The possibilities of a situation are expressed by "the noematic correlatives of our will awaiting us in the future" (Sartre 40).[5] Sartre emphasizes that the noema of a decision ranges over options, but of course it also ranges over times. Furthermore, the decision-making subject must be both factical and transcendent, both worldly and extra-worldly, to make a decision: "The personality must be made of clay, and I'm made of wind" (294). Put differently, the decision is about a quasi-actual world. It is neither hopelessly delayed in temporal indeterminacy, nor determinately synchronized with the future events. A decision's noemata are temporally variable, in-between imagined and actual time, virtual disjuncts in the will content.

Even if our will cannot control the future, as Sartre is about to argue, we can at least imagine future scenarios. On the one hand, of course, Sartre's interest is not in predictable risk assessment, or in Decision Theory's formulas for optimizing results of decisions under uncertainty. However, he is still interested in describing phenomenologically how we distinguish the willed future from the unwilled future. This gap between willing the future and controlling the future is the puzzle Sartre cares about.

Sometimes the temporal referent of a decision *is* fixed, as in the decision to wake up at a certain time tomorrow. Sometimes it is wide open, like when we decide that no matter what happens, we will regret nothing. Sometimes we will specific actions, but with vague temporal referents, for example, to live in a certain way once the war ends, whenever that is. These decision objects are not quite dormant, as we will them along the way. Sometimes, the whole business is complicated by stages and interacting alternatives.

Sartre's painful description of his decision to go on a diet ferrets out all the bad faith fits, starts, and downright postponements that are part of the decision from the start. For my part, I think these temporal rhythms inside what the decision means, make the decision genuinely futural, in and through its inevitable delays. Sartre draws a more negative conclusion from the same rhythms: namely, that the decision cannot carry its content into that future time when the decision is intended to bear fruit.

Part 1. Will and the future

The object of my will is distant from me by virtue of its position in time. Now freedom itself ... forbids you to will *against* time. You want to take such and such a step tomorrow. But who guarantees you against yourself? Tomorrow, your will of today will have fallen into the past, outside consciousness ... and you will be entirely free with respect to it: free to adopt it once again as your own or to commit yourself against it. One cannot swear an oath (*jurer*) either against oneself or against time. The pledge (*serment*) to oneself ... is a vain ... charm against his future freedom. Moreover, one swears only when he is quite aware there is a risk he may break his pledge ... What I will is my tomorrow's willing ... the duality willed/willing, but I precisely cannot will my subsequent willing ... I will a host of particular willings that threaten to escape me one by one. (Sartre 34–5/49–50)

If we consider merely the decision to *act* a certain way in the future, there does not seem to be any contradiction. Decisions are free, and once made, the subsequent actions are constrained. However, if we consider the decision to *decide* a certain way in the future, then there is a contradiction. As decisions are free, the subsequent decision must be just as free.[6] The contradiction is not just that we *will not to will tomorrow*, that is, that we will to bind tomorrow's actions. That would be bad enough. The bigger contradiction is that we try *to will tomorrow's will,* that is, that we will to bind tomorrow's freedom. We might think of this as a general theme in nineteenth century literature, from Melville to Zola to Bely: Did my own decision put me under the sign of fate, or is there still time to get out of it before it is too late? By Sartre's time, this issue is decided. There is time.

The first problem involves the doubling of will, identifying subjectivity and objectivity in the same act of will. The second problem involves the doubling of time, identifying present and future in the same act of will. To will an act at a distant time is to will (a) "against" time (as if not to allow time to pass), and (b) to will against oneself (not to allow one's freedom to exist later in time). Self-control itself is a self-contradictory notion, as it suggests "the image of a sturdy arm coming to check my arm. But I do not possess an inhibiting arm; I cannot personally erect barriers in myself between me and my possibles—that would be to abdicate my freedom and I cannot do it" (Sartre 125).

It is too bad, in a way. As Game Theory says, there is little point in making a promise to, or bargaining with, another person, if one does not have "the power to bind oneself."[7] However, if we don't, we don't.

It is true that "I am what I will" (41), but just in the sense that will negates what I am: I "will myself other" than I currently am; I will my will other than it is. Planning for the future is self-destructive in one sense, blocking freedom. Leaving the future open is self-destructive in a better sense, affirming my power of self-negation. Sartre rejects the humanist values of self-creation in Jaspers and de Beauvoir in favor of uplifting self-destruction. He does sometimes talk of the "obligation to remake the Self" (35), but as obligations go, it is not very encouraging.

De Beauvoir and Jaspers

In the terms of classical humanist existentialism, De Beauvoir and Jaspers are clearly on the side of both reason and transcendental phenomenology. De Beauvoir expresses the foundation of subjective temporality when she says that "will is developed in the course of time" (De Beauvoir 26). It is a condition for "decision" that one be "capable of recognizing himself in the past and seeing himself in the future," so as to "return" to the act of will and "prolong it indefinitely" (27).[8]

In addition, where De Beauvoir grounds decision using classical phenomenology of time, Jaspers grounds it with classical phenomenology of finitude. Freedom as a general idea means transcending my limits, Jaspers says. However, freedom as concrete resolution involves my freedom as a finite individual in particular circumstances. "Whenever I decide and act, I am not a totality but an ego in distinct given circumstances, in a situation that is objectively particular" (Jaspers 158). The individuality of the decision-maker grounds the connection between the decision the individual ego makes today and the decision the same individual faces in the future. It is true that I experience that "time is pressing," that I cannot fulfill all my potential, and hence that there is a degree of unfreedom. The existentialist ego is not a changeless substance, after all, but a self-affirming act. The fragility is precisely my experience of *my* freedom, especially if I understand my freedom as "choosing myself" (160), not just choosing things. "I am this free choice." For Jaspers, this demonstrates the opposite of what Sartre concludes. The absolute commitment to freedom of self does not lead to the unhinging of the self from the future; on the contrary, the freedom of the individual self projects the temporal life of the self into the future. "In regard to time ... I will unconditionally stick to [the decision to exist freely]". Otherwise, "I would destroy myself" (159). For Jaspers, it is practically a tautology to say that a free decision endures through time. The decision "sticks" because "resolution is the moment that can give my life a self-based continuity in the diffusion of existence" (159). This is orthodox existential humanism.

Jaspers quite reasonably wants to affirm both a de-objectified, infinite, self-choosing *and* the "press and stick" of time with durational continuity. Jaspers hits this two-sided nature of decisional time on the nose. Decision should both break and continue. It should be both virtual and real, both out-of-sequence and sequential, both beyond time and sticking to time. On the other hand, Jaspers' picture of temporality is one-dimensional and one-directional. The existential self, on his picture, has one future to go to, and one way to get there. It is just because of this that Sartre's view, no matter how unorthodox, unreasonable, and ego destroying, is more rigorous. The existential decision, with all its phenomenological force, may commit to the future; but the future of the ego will as a matter of fact have different choices when it gets there. There is plurality in the ego's ways of getting through time, and without an account of decisional plurality, and of the temporal plurality in the decisional future, Jaspers' affirmation of the committed self will be wishful thinking in the face of the gambler's next temptation. If Sartre's conclusion about decisional futures is extreme, Jaspers' is slack. I aim at a theory where plural, virtual time lines generate a way to keep the decision and its continuation, the real (future) and the virtual (multiple), together. How nice it will be if a theory of virtual time lines explains how decisions are both new and continued. However, in lieu of that, for the present, I think we have to be convinced by Sartre's challenge to the humanist existentialists' defense of the future of the will.

In short, willing a future will fails for four reasons: the self negates itself; time yields different situations; freedom is on its own; and decisions get reinterpreted. A decision is not a positive run-up to a positive image, but an under-determined project belonging to a subject who does not yet exist in a future that does not exist. An exercise in nihilation, the decision persists only "for as long as nihilation lasts, neither more nor less" (Sartre 41). We might even say a decision holds only as long as things remain under-determined.

We need not say that all decisions fail as a matter of fact, but only that the future is what resists. After all, the willingness to face obstacles is what makes a decision realistic. Future obstacles "distinguish the possible from what is, and envisage the possible beyond what is" (Sartre *BN* 37). On the one hand, the future does not exist when the decision is being made, so the future does not put up any resistance against the decision while it is made. In that sense, the future does not constrain what decisions we can make. On the other hand, the future when it arrives will be a stubbornly factual state of affairs, so by that time, it puts up too much resistance to the decision. These opposites—freedom and facticity—yield the same obstacle to the idea of a fulfillable decision, or a decision with a future. Will has no control. To put it precisely: We have will, but will is what we cannot do.

Sartre's formulation in *Being and Nothingness* is that there exists a "nothing which insinuates itself between motives and act" (*BN* 34). If there were no such obstacle of the future, then we *would* automatically act out our decisions, and we know we do not. However, we make this nothingness with our freedom, and it has an effect. This is not merely a normative rule against tying ourselves down.[9] We cannot help opening the future in the mode of decision.

Of course, we do have motives when we make a decision, and we do not forget afterwards that we had them. "[The former decision-maker] still thinks of himself as motivated; he believes in the effectiveness of this resolution," which is "always *there*" (*BN* 33). Sometimes, relying on a previous decision gives us an excuse for inattention later on, but sometimes it keeps us honest. When accused of failing to follow through on a decision, we cannot simply say that the decision belonged to an earlier time and is no longer relevant. But we can truthfully say that the decision has stopped being a subjective choice and has become an object, an object that we know we have to make a further subjective decision about. What is retained of an earlier decision is now merely the willed proposition in quotation marks, the proposition on the "plane of reflection" rather than on the "plane of action" (*BN* 37). Once a decision is no longer the concurrent act of deciding, it degenerates into memory or belief that one had once made a decision. It is modified from lived experience into a "detached" object for reflection. To put it simply, we know that we had made a decision, but we are no longer making it. The decision still appears afterwards, but "constitutes itself as ineffective." It is not just that that I am no longer my resolution; I am still that resolution, just in the mode of nonbeing.

A decision is thus weakened not by the amount of time passed, but by the way memory converts act to content. In Sartre's metaphors, the past is a "boneless phantom" that cannot communicate, or a "magic circle" that cannot protect us from the future, and leaves us "naked" "before temptation."

There are two forms of negation in consciousness responsible for decisions not surviving: (a) first, consciousness is "empty of all content" (*BN* 37), that is, it can take on any content, and (b) second, temporality nihilates the past of consciousness, so it is always empty again and again, and so always its own not-self in the form of its future.[10]

Sartre uses the vocabulary of in-itself and for-itself to define the ontology of time:

> The historian moves on three planes: that of the for-itself, where he tries to show how the decision (*décision*) appears to itself in the historical individual; that of the in-itself, where the decision is an absolute fact, temporal but undated; finally, that of the for-others, where the pure event is recaptured, dated and surpassed as being 'of the world' for other consciousnesses. (Sartre 300/364)

It is not quite that the in-itself is past, and the for-itself is future, or that the in-itself is factical and the for-itself is virtual. Sartre's model is more subtle; it is more like there are three kinds of virtual futures. The in-itself future is a plan disengaged from the subjectivity of a finite individual; the for-itself future is personal; the for-others future is being-in-the-world. The history of a decision is ontologically multileveled, as is its future (Sartre 301).

In summary, while conscious subjects act for-themselves with a future, objects, like the objective contents of decisions, are inoperative in-themselves. We could even say that the decision is "surpassed by the very fact that I am conscious *of* it" (Sartre *BN* 33). As soon as it is *fixed in* me, is becomes *unfixed for* me. The content of the decision ceases to be part of the subject pole, so I am not "subject to it" (*BN* 33). In other words, to be free, I have to will myself "*not to be* the past of good resolutions *which I am*". Of course, everyone would say this: We have to overcome the past if we want either to avoid repeating it, or to build upon it. If my past self wants to follow his decision, let *him* do it.

Nietzsche on promises

Nietzsche's *Genealogy of Morals* has a short, sharp analysis of the series of "paradoxical" moves that go into making a promise, or as he puts it, "to stand security for his own future" (Nietzsche

58).¹¹ First, to act because of a promise, one has to isolate one motive for action, and forget all the other desires and plans in one's mind, which might otherwise have counted as causes of the action. Second, of everything in the world contributing to an event, one has to isolate one's own action as its "necessary" cause, reducing everything else to the status of mere "chance." Third, one has to ignore all the conflicts and ambiguities in one's intentions, as if one had "decided" with "certainty" what one's goal was. Fourth, one has to presuppose that one "is able to calculate and compute" how to bring about a desired effect in the world; and furthermore, to presuppose that one has oneself "become calculable." One has to become an instrumental means to one's own end. For Nietzsche, these are conditions for believing that one can make and keep a promise: Suppressing most of one's own mind, bracketing most of the world's causality, and becoming computable. Each stage excludes a plurality, so as to focus on a single will, a single cause, a single goal, and a single computation. Each of these steps is unnatural and implausible on its face. However, Nietzsche presents it as a success story in the development of sovereign mind. If Sartre had operationalized decision in the same way, he would have concluded that decision-making is contradictory, that decisions never really take place in good faith, that we are in fact unable to "stand security for our own futures." What makes these steps absurd for Sartre, is the way they deliberately suppress plurality. For Nietzche, though, decision requires this extreme belief in singularity. So in spite of the rarity of those capable of it, he regards it as forcing a new figure onto the scene, whom we might call the over-decision-maker. When, later, we pursue an alternative both to Sartre's skepticism and to Nietzsche's exceptionalism, we will need to find a mechanism of decision-making that does not involve Nietzsche's stages for denying plurality.

It might look as though there are no future effects of the past for Sartre, or even that the past is nonexistent. As the past is nonbinding, he goes so far as to say that "it is of no importance to have this or that past" (Sartre 335). In order "for [the past] to exist, we have to throw ourselves through it towards a certain future" (335). On the other end, it looks as though there is no future in Sartre, since it is so unbound. However, Sartre's intent is that the future is laid out by options precisely because it is unbound. There is a future will; there just is no causality of will. There is obviously a paradox in this. It might seem as though either of the two common stories about decision—that it inaugurates a radically new future, or that it radically breaks with the past—makes decision temporal in its own way. But for decision to be thoroughly in time, its future has to feed back into it. Yet Sartre clearly has a point in emphasizing that the will in the present does not determine the future will. So whatever kind of feedback the future will has on the present will, and vice versa, it will have to be conducted along ambiguous lines, along many different lines, some of which describe what the will is doing and others of which say nothing about what the will is doing. This is why I think a multilayer theory of time will be necessary for a good theory of decisional futures: Decisions will have determinate effects along some of its futures, and not along others, and all of those futures will constitute the meaning of the decision itself.

The future looks especially unstable during the Phony War, when plans are on hold. But instability is a universal condition of all futures, and because the future is unstable, so is the present. Indeed, the intended future is a condition for the present, which would otherwise end in the now. The future can be a "reprieve" (107) for the will, even if it cannot be a realization of the decision. Sartre's spin on Husserl's "epoché" is that the decision brackets the certainty of the world. The decision is indexed not to a future state of the world, but to a virtual time in "flight" from the world (112). This is after all the ethics of ambiguity. The future destroys what one willed just by affirming that will. It is not weakness of will, but strength of will, that makes decision fragile.

I am not primarily interested in the sources and cures of weakness of will, the Aristotelian question, but in how decisions view the future. But of course, these two questions may be related. Time passage might well generally be a source of will decay. Conversely, if decisions look forward to multiple futures, not just to one future, that might multiply the number of results that count as its enactment, and therefore reduce the number of cases of the weakness of will. There are, after all, fewer failures to do what one said when more things count as doing what one said. However, time passage is not always relevant to the weakness of will. In Aristotle's (*Nicomachean Ethics* Book vii: Chapter 3) syllogistic model of decision ("I should eat less fattening foods; Bread is a fattening food; Therefore I should eat less bread"), the major premise "I should eat less fattening food" need not alter over time, nor should the minor premise "Bread is a fattening food." Aristotle would agree that the will acts anew each time temptation arises, and Aristotelian wisdom does match judgment to changing circumstances. Still, the syllogistic interpretation of will does not focus on temporal projection in decision content, or on time as a source of will decay, and that is my interest.

In fact, if time passage were by itself a source of weakness of will, it would provide a transcendental excuse for failure,[12] a kind of willful fatalism, like carrying a poison pellet to kill oneself the instant things go bad (324). Sartre wants to reject escapism, to accept freedom's negativity. Moreover, he has a resource for doing so, which is to treat the future not as annihilation, but as a plurality. Indeed, these two motifs—annihilation versus multiplied otherness—define two versions of dialectic.

When Sartre takes the latter, pluralist route, which he does not always do, he says that today's will covers a "host of particular" wills for future times. Consciousness throws itself into the world "at every instant," "refracted through the diversity of the world", signifying "a multitude of exigencies" (37). The very manifold in the world at present implies a manifold of futures. Each decision is many decisions. On this version, it is the multiplicity of potentialities inherent in a decision that makes the decision impossible to carry out in one actual future. "To change one of my possibilities is to change all at the same time" (41). Noetically, each act of will is one act at one instant, but noematically, the options are many, as the world is many. Moreover, a plurality of optional noemata corresponds to a plurality of temporal noemata.

Again, the plurality of the future is what is at stake in the decision in the present. "All such options are simultaneously present to us" (41). We might wonder whether it is really consistent that there be one will per instant although there are many options per instant, and although the future we project into is many. Should there not be many wills per instant to match the option count? Not all options can be thematized, of course, as we cannot imagine all future possibilities. So it is fair to say that the particulars of a decision exceed the "situation" of the decision. Nevertheless, the many particulars have to be thematized a bit in the decision unity, precisely because we choose among them.

The plurality of exigencies can take various patterns. Some decisions are temporally simple, and almost immediately find fulfillment, for example, lighting a cigarette as soon as I think of it. Most are temporally more complex, because ontologically, there has to be some kind of temporal interval between decision and act. We could even say, as Jankélévich puts it (channeling Bergson), that the future is precisely the "temporality where we cannot do what we want."[13] Sartre says of deciding to write a book, for example: I can do some of it immediately, but by necessity I can also turn away, or be turned away, during the process. "The permanent possibility of abandoning the book is the very condition of the possibility of writing it" (37). Eventually, the book takes on a "relatively independent" future structure. In addition, there are still "more distant ends": The end of the work, further books and readers, each carrying anxiety on different scales. "Is it necessary to write this work?" "Is it opportune to write it *at this moment*?" (36).

Difficulties arise not so much from temporal gaps or delays in carrying out a decision, as from the mixed rhythms involved in carrying it out. Sartre decides, for example, to write to de Beauvoir every third day, and to his mother every second. Moreover, while it takes at least three days for the letters to be received, he finds that he imagines them received immediately (65–6). It is painful to be out of step with our decisions, so our imagination constructs overlapping artifactual simultaneities of present and future. That is the rule.

Part 2. The diet: Breakfast and time

What I should like to signal today ... is the *way* I'm faithful to a decision (*décision*) once it's taken. For example, I can by and large say that I've been faithful, yesterday and today, to my decision to take just one meal daily and not to drink wine or eat bread. But seen close up, this triumph breaks down into little specific defeats, just as battles if they're seen close up are always defeats for the victor. In the first place, once the decision was taken I regretted it and added a postscript: 'Well, actually, for my breakfast I shall eat bread after all.' Not so much because I considered this unimportant, but because, after a rapid inspection of my possibles, I saw I couldn't help doing so. When one wishes to take any decision, one takes a good look all around and inspects one's possibles. There are some which are hard as rocks and must be skirted, while others form soft, jelly-like masses and that's where one must direct one's efforts. (Sartre 122/155)

For Sartre, breaking a diet has little to do with hunger, or overdeveloped tastes, or underdeveloped self-image, but mostly to do with time. The usual ways of keeping a diet are in the worst faith, indulging in rewards, competitions, or identification with body image.

It sounds funny for a soldier to worry about his figure. But after all, the decision to diet is notoriously difficult for anyone. It is to be expected that popular diet advice emphasizes the problem of "sticking to it." Fitnessmagazine.com offers ten ways to recognize "The Moment" when a dieter is most vulnerable to temptation, like when "Your TV cuts to a commercial." Recognize the moment and control time. Fit-not-fat.com suggests you make ten reminder cards listing your motives for going on the diet. Return to the origin and control time. Times-of-India.com is more fatalistic; its best advise is ten ways to "limit the damage."

Sartre says that he "distrusts" the seriousness of actions. "I'm recounting all this ... feeling rather ridiculous" (Sartre 123). In fact, decisions are normally regretted. Post-decision regret does not necessarily lead to its cancellation, but it does generate postscripts and provisos to the decision. Just as battles have aftershocks, so contracts, including contracts with oneself, add riders, demanding a bit more or less, while retaining the agreement's general force. This is part of what Sartre means by having "a good look around" at one's possibles. We can never anticipate all the provisos that we will want to make after deciding, so providing the time for regret and revision is part of the process. For some decisions, enactment turns out to be just too hard. Maybe we should have known that in advance, but gifter's remorse is inevitable. Moreover, we can still work on the soft parts. Sartre measures the hard and soft elements of a decision by the quantity of "effort" (122) needed to enact them.

William James

Effort Psychology sounds more like William James than like existentialism. Sartre usually rejects the formula that decisions fail when temptations require more "effort" to resist than the agent has to give. It might seem plausible that temptation conflicts with past resolution, as desire holds an "inner debate" with reason. However, Sartre figures we rarely "weigh motives and incentives before deciding" (Sartre *BN* 32–3; compare with Hegel in Chapter 8). In brief, the theory that intensity of motive generates degree of effort has four problems. First, it is a false description of inner experience. Second, it dogmatically assumes a materialist metaphysics of psychic "forces," whether conscious or unconscious. Third, it falsely assumes that people's choices conform to an economy of incentives. And fourth, it assumes a linear time line: Prior deliberation, the moment of decision, and subsequent action. Effort Psychology does not have a way of retroactively including the postscripts of a decision into the meaning of the decision itself.

Still, Effort Psychology has some interesting ideas that could add to a phenomenology of decision.[14] In James, for example, the picture of decisional "effort" is less like pushing with a strong arm and more like filling up a balloon. "Decision" means letting a single idea "fill" the mind (James *Psychology* 560). Even right and rational decisions, James says, can get "crowded out of sight" by excuses. James' example involves abstaining from drink rather than food. When James asks, "How many excuses does the drunkard find?," he runs through a fantastically long and imaginative list. However, James figures, if a repentant drunkard could only fill his "attention" with the idea that he is a drunkard, and "keep the right name ['drunkard'] unwaveringly present to his mind," he might be able to stick to his decision to get sober. James is as attuned to temptations and excuses as Sartre is, but there is no futural content to James' account, so it is hard to see how James' "unwavering present" can ward off the promise of future enjoyment that excuses gamble on.

In other writings, James experimented with more subtle accounts of the decisional present. In his "Draft on Brain Procedures and Feelings,"[15] he says that when we are deliberating among alternatives, we experience suspension and oscillation. Then we experience "a sort of click" (James *Brain* 255), after which "the act either immediately ensues, or the steady intention calmly waits for its opportunity." We call this click: "decision" (*Brain* 255). What really happens, he says, is that the period of suspense consists of "simultaneous" brain waves interfering with each other, and this is followed by one wave "breaking through the dam", so that "the total nervous movement changes its form from elastic oscillation to definite advance" (*Brain* 256). That explains the click. In this text, which essentially rejects Effort Psychology in favor of an interaction of psychic forces, James gives a more layered account of the decisional present, which might be more useful for Sartre's ambiguity version of phenomenology. But if anything, this version of psychic decision is even less futural. In addition, it does not leave much room for Sartre's insistence that the future we will come to will be a hard obstacle to the future we intended.

In *The Will to Believe*,[16] James gives advice about the timing of decisions. Do not make "decisions for the mere sake of deciding promptly" (James *Will* 20). On the other hand, do not put off decisions as if they do not matter at all, or as if permanent skepticism could replace decision. Decision is not for speed, or for resolution at all costs, but for truth.

However we measure degrees of hardness, breakfast is a rock. Sartre cannot help eating breakfast. He "doesn't mind" missing lunch. ("Minding" would be an interesting category for the philosophy of mind.) He says, and we might as well believe him, that he does not mind "eating salad with no bread [for lunch]; or actually fasting for a day or two" (Sartre 122). However he likes breakfast, for three reasons. First, breakfast makes him feel "poetic and fragrant," not to mention "lucky" (122). Second, he likes to be alone at breakfast time, but he needs a substitute for the other person, and that substitute is buttered bread. Third, and most important, breakfast helps him *not* to wake up too fast (to "prolong [*prolonger*] the morning," 123); and prolongation helps him think. In fact, delay generally is hard to sacrifice, because delay for a philosopher is a value in its own right.[17] (While he wants to delay the day's business, he does not want to delay having breakfast—he wakes up his fellow soldiers earlier than even soldiers need to get up, at 6:00 a.m., just to get to breakfast—he wants the delay *of* breakfast, not a delay *in* breakfast.) Breakfast delays the other's intrusion, and also delays lunch, the metaphorical "meat" of the day's menu. Delay keeps the future from arriving too early (123). One might think that a delay could keep the decision from dissipating. However, delay comes too late for the decision, which is defunct already at dawn.

Now, there is a way to get time onto the side of decision, if the enactment can be forced to yield something very much like immediate gratification. Sartre knows of himself that "I'm always in a hurry to see the effects of my diets. So I always choose the extreme course and prefer to torture myself a bit, because then it seems to me I can *feel* the progress of my slimming through the protests of my stomach" (123). Indeed, quick results give a secondary positive feedback for not feeding oneself, namely that "if I crack down on myself a bit roughly I have the impression of being my own master, hence free" (123).

Metaphysically, this is a slick move. If the dieter can force the future to happen right away, then a decision could engage freedom by brute force. Of course, even the short run experience has to be long enough to lose a bit of weight, say one month. However, the month, the shortest time that can yield immediate gratification, is already too long. Sartre recounts that:

> a month of constraint, with me looking at myself every day in the mirror to see my progress and weighing myself on those automatic scales which chemists place in front of their doors, and then—the target reached, or reckoned to be—I go back to living as I please, I no longer watch myself, I grow fat, until the day I begin looking anxiously at my belly again and reflecting on what measures must be taken to deflate it. (Sartre 124)

These repeated cycles, "in fits and starts," do not cancel the positive result of reading the future quickly one effect at a time. One can lose some weight in a month, after all. However it would be too good to be true if impatience were a virtue for decision keeping.[18] Sartre's ruthlessly critical friends tell him that he is the sort of bad faith masochist who gets pleasure out of forcing himself to do what he does not like (Sartre 124), and Sartre is too honest to deny it. It is a corollary that Sartre gets pleasure out of telling himself what he does not want to hear, sadistically demeaning the masochist within. He tells people he is on a diet, not to boast, but to "burn his bridges and commit myself more fully" so he cannot get out of it without shame (124), and will have to "stick to these decisions to the bitter end" (124). However, he knows they will make fun of him anyways, and then he gets pleasure being made fun of, as that shows how good and honest he is after all. There is a vicious circle of pride and shame.

What conclusion should we draw from these self-critical insights? Do the backhanded gratifications motivate staying on the diet, or are they gratifying whether he stays on the diet or not, or are they anti-gratifications that weaken the diet, or anti-gratifications that motivate staying on the diet? The impossibility of judging which is which will lead to Sartre's conclusion.

Eventually, the same cycle of mastery and self-indulgence that makes it pleasant to get thinner, leads to relapse. Sartre says he has a "holy terror of all those fellows who decide at three-monthly intervals to stop smoking, keep it up for a day or two with a hell of a struggle, and then give in and start smoking again" (124). The whole point of making a decision is to stand like an overman above such weaklings. But Sartre's horror of them does not make his results any better. "My way of standing fast is not noticeably different from their way of giving in" (124). Their indecision is in its own way a decision that lasts a little while, and his decision is its own way of not deciding forever. Existential commitment is never more than this.

However, we have not yet examined the concrete details. The particulars of breakfast are important—for example, that breakfast must consist of bread or croissants, not fruit—as is the specific café where one takes it. Obviously, the details of Sartre's case are not universal, and his particular decision anxiety might be abnormal. However, there are obstacles and temporal vicissitudes for any decision, and in that light, Sartre's diet is universal.

At any rate, Sartre thinks the most dangerous obstacles to decision are in the particulars. He is interested in the particulars not merely because individual people cannot break their personal habits. He is interested in repetition under different factual conditions. The immediate motives to diet are after all different each time. The first time he decided to diet, he recounts, was because he felt "the horror of getting fat" (123). Another time, it was because De Beauvoir pleaded with him. We can generalize, and say that decisions normally have previous installments, often associated with love affairs and other people's judgments, and with specific places—the Café de la Poste on Boulevard Rochechouart, and the Dôme of course. Sartre names his favorite diet crushing restaurants because he is writing a diary, but I name them here to make the reader hungry, so the idea of a diet will be appropriately bothersome. However, as telling as they are for concrete description, backgrounds do not exhaust a decision, nor do prior occasions, or previous defeats; obviously Sartre wants to diet again.

It is here that Sartre comes closest to citing his motives, although what he describes is more a history of images than of reasons.

> The horror of growing fat came upon me quite late. When I got back from Germany [where he was studying Heidegger] I was a real little Buddha [he means his stomach]. Guille used to grasp handfuls of my paunch through my sweater to show Mme Morel I had enough and to spare, and I used to laugh easily about it—it didn't bother me especially being fat. But when I got to know O., I conceived a horror of fat people and began to dread the idea of becoming a bald little fatty. To tell the truth, I would have a slight tendency in this direction if I didn't watch myself. But precisely, I'm incapable of watching myself. That Lady [almost everyone involved either in promoting or undermining Sartre's diet are women—the men on the scene, soldiers and doctors, either do not care, or are fellow decision-breakers] and le Castor [de Beauvoir] have often pleaded with me to follow some gentle permanent diet. But I'm absolutely unable to keep a check on myself without lapses." (Sartre 123)

Of course, we should not assume that permanent change cannot be decided. However, permanent change is more like transformation or reformation, and less like decision.

In summary, almost everything about the breakfast decision—delay, repetition, particularity, mastery and shame, short- and long-term outcome assessment—leads to decision-deformation. This is not very encouraging. However, for Sartre, the value of decision-making is only marginally connected with success. It must be about outcomes at least a bit, and certainly values are at stake, but for Sartre, outcome assessment does not shape values. His appeal to cracking down on the body suggests deontology over pleasure, will over utility, process over ends, masochism over power. In any case, the decision transgresses over actuality, and this provides a pleasurable outcome in the interim. It is not that the pleasure is in deciding whether one slims down. It is that the pain gives the feeling of slimming. The decision, to be worth making, should be extreme, even if that "defeats" it. The decision I cannot maintain makes it mine. Moreover, if only for that reason alone, the dieter's decision is not over yet.

Part 3. The diet: Lunch and the other

We learn only in an aside that Sartre did not eat the croissant for breakfast. Another aside: He drank a few drops of wine in the morning. He decides on no bread or wine with lunch. Somehow, he thinks doing this will lose weight. There is no mention of the lunch menu (or of exercise). As with breakfast, he lunches alone. There is no *Mitsein* at the table. After all, I can no more swallow the other person's swallow for them than I can die their death.

Sartre introduces lunchtime with the phrase, "So there I was having lunch" (125). He uses this phrase "*me voilà donc*" four times on the page, as if he is surprised to find himself eating. (A phenomenology of decision should devote more time to the logic of "*voilà.*")

The decision to diet is still "superficial" until a few things have actually been renounced. So at first he forgets it, in the "conviction" of lunch as usual. As a general rule, "the restaurant at midday was a perfectly round, hard plenum, closed upon itself, where I had my place" (125). This much is true of every decision; there is a tendency to forget one made it. The decision violates the world around us and the facts of the matter, and appears as an "objective impossibility." Rather than the world appearing as the obstacle to surmount in favor of the decision, the decision appears as the obstacle to real life. In fact, there are times when it is hard to tell if Sartre is describing his temptation to break the diet, or the temptation to go back *on* the diet. Either way, with any obstacle, one's first thought is to "get around" it. Of course, we cannot get around having a future, but the future is all about getting around things. By analogy, Sartre says, if I want to meet A, and A is out of the office, I can try to get in touch with him at home. So if my resolution is unpleasant, I try to find another way to eat lunch, to get around my own memories and engagements. The way around the obstacle of this decision is to eat lunch anyways. It is the decision that is in bad faith, I can always say, since it presumes a controllable self. In the same way that I was free to decide not to, I am now free to eat the bread in spite of my decision (125). The decision never became real, it was only ever in my mind. Existentialism itself is the excuse.

Let us think about this as a practical problem. It is true that I cannot make the diet objectively real, but do I really have to break every decision on existential principles? If I cannot uphold it, perhaps I can at least "refrain" for a while from breaking it, "'to put off' (*différer de*) canceling the decision, to remain in suspense," to delay disobedience, to keep the decision operating precisely through indecision, in a state of "perpetuated hesitation." Of course, a hesitating expression of a decision does not always signal the intent to remain hesitant forever. As Silke Panse shows,[19] drawn out decisions in TV judgment shows like *Masterchef* and *America's Got Talent* turn a pregnant pause before a decision is announced into a ruse for the simplest declarative decision (Panse 58).

Perhaps we can use to our advantage the vicious cycles of collapse and reform (Sartre 125). This is to be sure a "sly" and "flabby" sort of decision. However, one can talk oneself into not quite abandoning the decision just by asking, "Oh well, is it really worthwhile eating bread?" How long can indecision save you from the bread? Only "until (*jusqu'à ce que*) the world changes": Only until the next time (125).

The "until" clause is a problem for any decision, existentialist or regular. How much does the world have to change before we can fairly stop trying to adapt a decision? This is analogous to asking how much perception is needed to falsify an assertion. Epistemologists sometimes say that such "how much" judgments are made pragmatically in context, but that concedes that there are no philosophical concepts in the relation between will and the changing world: like virtual time lines. In any case, Sartre's answer is minimalist: For him, very small changes call for brand new choices. It is, for Sartre, valid to prefer even a minimum amount of freedom over a great likely outcome.

Articulating this trade-off hints at the economics of Decision Theory. Phenomenology, although obviously not committed to utility maximization, does have a description of those experiences where we compare values. It is not out of place when Sartre asks, "Will [eating bread] give me enough pleasure [perhaps he should have said—will give me enough freedom] for me not to regret having broken my pledge?" (125). In fact, Decision Theory offers two schemes that Sartre might have used. First, Decision Theory distinguishes sequential, repeated, and iterative decisions (see Chapter 6). A sequential decision is made in steps. For example, a person can either decide at one time what to order for appetizer and entrée, or can sequentially decide on the entrée only after finishing the appetizer. A repeated decision is like choosing among the same appetizers every day—one-step decisions made on multiple occasions. An iterated decision plays out complex decision sequences on multiple occasions, learning from mistakes, experimenting with new strategies. It might have helped Sartre to consider the way decisions can be broken down into subgames or branching nodes. In addition, the competitive aspect of typical Game Theory decisions raises the second issue largely missing from Sartre, namely intersubjective decisions: "tit-for-tat" responses and negotiations. Sartre does have his own limited account of intersubjectivity in decision. For there remains still a missing element.

"There remains the wine" (125). In this case, the restaurant's wine "isn't very good: It has a pink cloudy colour that doesn't tempt me, and a sugary acidity that reminds one of apples more than grapes" (125). Sartre has a nice reason not to drink it. However:

> There now occurred the kind of event one should always reckon with, the kind one never thinks about and whose specific character is to lead you into transgression by offering you a ready-made excuse ... Before I'd ordered anything, the waitress went off with a smile to fill a flask [of wine] from the barrel and placed it on my table, as if saying 'You see, I know your tastes.' She looked happy at being acquainted with her customers' preferences, and I didn't have the heart to disabuse her. (Sartre 125)

Sartre therefore also has a nice reason to drink it. Both the ethics of intersubjectivity, and the greatest happiness principle, even simple "human respect" (125), are all on the side of breaking the diet. One cannot make one's decisions in a vacuum, in a "still-life" without other people (*monde morte*, 126). Even when you eat alone, intersubjectivity is on the table. In *Being and Nothingness*, Sartre describes the man in the park whose world-interpretations are interrupted as soon as he notices another person in the distance; it is as if his own worldview is "sucked down a drainhole" that lays in wait precisely where the other person is standing (Sartre *BN* 256). Here, in the flask, Sartre's image is less of draining out than pouring in, but one person's will is still drowned out by the mere consciousness of the other. It is not just a matter of my decision's unintended consequences for other people. It is a question of what the decision's content was about: "saddening a waitress's heart—that wasn't part of my contract" (126). The decision conditions were not "realized" because other people had not been taken into effect, so the decision is null. Decisions are always null as soon as other people enter the room.

In very general terms, we could even describe intersubjectivity as a whole, the recognition that the other is other, not yet in our grasp, as a generalized "hesitation."[20] No doubt, this is a rather personal view of intersubjectivity. In a political setting, the intersubjectivity issues might be different. Adorno's view,[21] for example, is that Sartre's thesis that decisions have no future is only correct for certain types of society, namely bureaucratic and fascist states, where individual decision-makers are alienated from, and disregarded by, their societies. Under such conditions, there is no way that an individual's decisions can have continuous, let along lifelong, effects in the social world. As Sartre lives in such a society, he thinks human choices can only have the form of "jolts", rather than belonging to causal chains (Adorno 226–7). This is why Sartre "clings to the decision alone" rather than to revolutionary social action. Sartre's limited range does not represent the "*condition humaine*" (226n); in a different sort of society, decisions might have futures. However, Adorno is not generally optimistic about the possibility of truly free political life (i.e., of a positive and not a negative dialectic), where individual decisions would be continuous with social causality. In addition, as long as a better world is delayed, Adorno has to agree that Sartre's descriptions are correct so far.

Once we raise the issue of cultural difference, there are many other issues we could choose to deal with as well. For one thing, cultural differences in narrative will have an effect on a theory of time and decision. Lyotard[22] argues, for example, that it is primarily only in the West that narratives tend to accumulate content. In traditional cultures, "nothing gets accumulated, that is, the narratives must be repeated all the time because they are forgotten all the time" (Lyotard 34). I do not know if this is true, but the idea is that while time and accumulation (or anti-accumulation, in the case of dieting) are generally connected in the West, they do not necessarily go hand in hand. Such an asymmetry of accumulation and time might require a slightly different formulation of Sartre's asymmetry between decisions and the future. Instead of saying that decisions have no future, we might instead want to say that decisions do have a future, in that they need to be repeated, but that they do not accumulate over time. Or that they constitute a time of relayed repetitions, but a time without an open future.

Mbiti

I have been presuming that it is possible to have an experience directed toward the future at various distances, but we cannot take this for granted. John Mbiti[23] argues that African cultures and languages do not recognize a distant future. This is a controversial thesis, and other African philosophers have replied that African languages do recognize a distant future. I cannot verify Mbiti's claims regarding the Kikamba and Gikuyu languages, but it is interesting to consider what time looks like without a distant future.

Mbiti's ontological premise is that time is grounded in events, particularly human events, and that events naturally occur in cycles. In agricultural societies, events are structured around times of day and seasons of the year. Each event takes a certain amount of time, ends at a certain stage of the cycle, and is repeated as the cycle returns. Mbiti acknowledges that urban industrial societies break this chain, and at some points, he argues that African societies need to change their traditional ontological concepts to succeed in the mechanized world. However, he also suggests that the concepts of time in traditional society reflect better what time really is.

The key point about cyclical time is that events renew the past. Obviously, some events may occur that have not yet occurred, but the future consists primarily in completing an event that is already largely past, and then starting again. The kind of time that continues the past is "potential time," that is, time grounded in events; the kind of time that aims at an imagined future unconnected to what is actually happening now, is an abstraction, a distortion. In this picture, Mbiti says, time has "a long *past*, a *present*, and virtually *no future*" (Mbiti 21). Mbiti favors the paradoxical formulation that "*actual time* is therefore what is present and what is past. It moves 'backward' rather than 'forward'" (23). In Kiswahili (28–9), "zamani" is the long-distance past, which grows increasingly in length and content as long as humans and societies live and are remembered. "Sasa" is the dynamic present, in which an event is acted out. It is a short, but expanding and contracting present. The dynamic present is not an instant, but includes the momentum of an event that began in the near past, and continues into the near future. There is, in this schema, a minimal conception of the future, only insofar as an event has already been decided upon and is continuing to guide actions and perceptions. The future does not last a long time.

In one sense, this is phenomenologically opposed to Sartre, as here a decision is presumed to carry a person and a society into the future. However in another way, Mbiti's picture shares with Sartre's the idea that one cannot just decide what will happen in the future if one is not in the process of carrying it out while making the decision. In both cases, imagining oneself doing something in future time does not posit the future as a real piece of time. That sort of imagination is more like storytelling than deciding. A genuine decision extends only as far as the present extends.

These cross-cultural variations may require revisions in Sartre's formulations, and they add levels of intersubjectivity to what an individual can do with her own decisions. However, we still want to see what happens when an individual's decisions encounter another individual's consciousness.

Sartre's focus is on a rather lonely individual encountering another in a public place, and in that setting, Sartre's feeling is that he "hadn't made preparations" (Sartre 126) for the other's role. The other, from the perspective of my consciousness, is "virgin territory," which deliberations cannot have included in advance (126). Sartre figures he has to "postpone deciding until later (*remis à plus tard de décider)*" (126). "I'll start my diet tomorrow, today it's impossible, nobody's obliged to do what's impossible" (126).

However, as he has good reasons both to follow through on the diet, and to break it, there is momentum for compromise. He finds an "expedient":

> I would pour a little wine into my glass so that the waitress, seeing the flask half-empty, would simply think I hadn't been very thirsty, and wouldn't notice that the contents of the glass corresponded precisely to what was missing from the bottle. In addition, to complete the illusion, I would take just a sip of it. (Sartre 126)

We can analyze this compromise under several parameters. (a) Temporally: Compromise is "suited to the present situation," that is, to the compromise between the diet on the one hand and respect for the other's feelings on the other hand. However, in conflict with the present, violating the vow makes him tantamount to "the merry toper (*joyeux buveur*) pouring himself a beakerful," abandoning the present, reverting to his past, pre-decision, overindulgence. (b) Psychologically: On the one hand, the compromise is in good faith, as it is done for the other's sake; on the other hand, he knows his good faith is in bad faith, and is afraid of "being drawn too far." (c) In terms of pleasure: On the one hand, the wine loses its appeal, since he knows he is failing by drinking it; on the other hand, there is the "furtive pleasure" (Sartre 127), drinking supposedly for morality's sake, then abruptly setting the wine back down on the table, in the "fear" of getting caught. Sartre uses a horrible analogy. He feels he is "reminiscent of a doctor who 'takes advantage' of auscultating a beautiful patient …, who enjoys her via his fingers without halting his professional explorations" (127). Sartre does not say that he is abusing the waitress by taking his pleasure from her innocent gesture, but it may be that ethics based on the supposed pleasures and pains of others is often an excuse for harming them. He is not quite using *her* as a means to an end; he treats *her end* as a means to an end. This might make an interesting rule in ethics—there are times when considering others as ends interferes with morality. (d) "Symbolically": The expedient allows him to "mime" (126) the very thing he is not doing. He drinks, but only pretendingly; he diets, but not really. He doubles down both ways, he decides in the form of nondecision.

In the final analysis, strangely, "from the outside, in spite of everything, it's a *successful* act" (127). Sartre will lose a bit of weight; he drank "only a tenth of a glass instead of the two full glasses he was used to. He said he wouldn't eat bread, and he hasn't" (127).

Here is Sartre's assessment.

> The victory is a Pyrrhic one. I shall have another five or six [days cutting back on wine] like that, then the *habit* of eating meals without wine will come to me, and in that way I shall *in effect* have kept my word. But when all this indecision [*veuleries*] disappears, consciousness too will disappear and the act will become automatic. That's why, when I receive any kind of praise, I always have the impression it's really directed at somebody else. There is no act without secret weakness … That's what's called having will-power. You can see how much it's worth. (Sartre 127/161)

What is true "in effect" is false in cause. The belated role of habit for Sartre is not the everydayness that prevents the decision from changing the norm; on the contrary, habit is what allows him to carry out the decision successfully, but at the price of making it something other than free decision. Success as such is an encomium to bad faith. The decision succeeds as a past, but not as a future. The outer world, the body, accumulates progress; the inner remains unfulfilled. The outer is like "somebody else." Success is somebody else. Or to be precise, success is me thinking that somebody else is thinking of me as somebody else. I evaluate my decisions through a keyhole.

In some ways, this is also Sartre's view of soldiers in the war diaries. Only those who do not receive a future deserve to have one. Only losers deserve to win. Moreover, this is Sartre's last word on decisions and the future: Insofar as a decision affects the future, it ceases to have a futural noema. This is why the decision cannot carry into tomorrow in the form of a decision. This is why Sartre drinks wine with lunch.

The journal entries on following days do not concern the dieter, but I am going to read some of them in the form of two postscripts.

Postscript 1: Decisions are enacted otherwise than originally intended

A diet could always lead to exercise instead of weight loss. "'All that I used to want in my youth, I have had—but not in the way I wanted it' ... Our great hopes are dead, and ... it's them we look at through our success" (Sartre 198). Of course, if we *now* "compare" what we have with what we wanted, we find a big difference. However, that comes from identifying the past desire as present, as if we were still in the past deciding. As decision occurs in a *changing* situation, it *should* get enacted differently. In addition, as there are layers of cause, effect, and signification (294) in any decision or event—military and economic layers, character traits of the agents, their physical and psychic disabilities, and literary styles—it may be undecidable (without engaging Nietsche's forced simplification procedure) which among those events that follow a decision are the events that enact it. The virtuality of the decisional future is tied to the unsynthesizable layering of historical explanations.[24] Decision has authentic indecisiveness inside its noema. It is not that authenticity is unrealizable, but authenticity affirms the unrealizable.

The details of this theory would be a mess if we could not interpret the future as virtual time. Should we say the actual moment of decision is in time, but its underdetermined enactment is not in time? That events are in time and decisions are not? Or that their enactment has a temporal character, but not in the actual future? Or that factical occurrences have the status of being past, even if they have not taken place yet, while free decisions have the status of being future, even if they have already taken place? Or that one and the same event subsists both continuously and discontinuously? Or that the future referent and the future enactment are different kinds of future? I am not sure which to pick. In any case, there have to be two types of time: dated actual time, and branching virtual time.

Postscript 2: **The person addressed by the decision is different by the time it is enacted**

Of course, when a person promises to love us, we want him or her to love freely, not to "act out the determinism of passion" (258). We cannot consistently want our loved one to have "pure fidelity to the pledge of fidelity we have just torn from her" (257–8). We do not want our lover bound by decision. We want the decision to be constantly futural, in the act of deciding, never at the stage of having been carried out. And surely she must want our receptivity to be just as indecisive, and why would we refuse?

One of Sartre's nicest ideas is that in decision, one makes an appointment with oneself for a later date. In the *War Diaries*, he speaks of "the rendezvous I'd fixed with myself for after the war" (222). In *Being and Nothingness*, the anguish is that I will miss the rendezvous, that I will get to that later time, "search for myself", but not see myself there, or I will be there, but I will not go there to find myself "… on the other side of that hour, of that day, or of that month. Anguish is the fear of not finding myself at that appointment, of no longer even wishing to bring myself there" (Sartre *BN* 32). We say loosely that I will be a different person by then, but perhaps it is better to say that I will still be myself in the future, but in a different possible world. Or, since my spin on possible worlds is that the same world contains different possible time lines, the way I would express it matches, though with a twist, what common sense says: Namely, that I will be myself in the future in a different time.

The strong version is that "I am already there in the future" (*BN* 31), but "the future which I am remains out of reach" (*BN* 36). Can this literally be true, not just an expression of anxiety, that I am out of reach of myself because I exist now in another time than the present? That I will be in the future twice: Once in the future as my future self, and once in the future as the self I am at the present? Can another time exist at this very time? Can this time exist again at a later time? This is an old problem, from Aristotle to Derrida: At the time when one now passes into the next, is there a bit of time when now and future, as distinct times, both exist simultaneously?[25] Is the future an existing but underdetermined property of the present?

Sartre may not require quite this much ontology. Perhaps he should stick with the phenomenological version: "I await myself in the future where I 'make an appointment with myself …' " (*BN* 36). It is an appointment (a) with time, (b) with a self whose desires will have changed, and (c) with the noema, as "it is only the distant and presupposed meaning of the act which reveals my possibilities to me …" (*BN* 37).

Suppose I do meet myself in the future. How will I recognize me? I will have to "discover the essence of *what I have been*" (*BN* 37). I will have to wake up one morning thinking, "I have been 'wanting to write" this book. My future's past, that is to say my present, will for the future be a means to an end. Moreover, such temporal mirrors multiply with reverse rendezvous. When he was a youth, Sartre recalls, "I'd have liked to be sure of becoming a great man later on, so as to be able to live my youth as a great man's youth." In that reversal, he will even now ("now," meaning his childhood, which he, the 34-year old journal-writing Sartre, still before he did become a great man, is recalling) "go for a little stroll in the future" (Sartre 74).

In summary, I am the present one who intends nothing other than to affect the future, and I know that in the future I may read it differently, but that is exactly what I am about when I decide. I am the future one who is not obliged to be it. "I *am* not the self which

I will be" (31). Of course, to call it "my future" means precisely that it is not necessarily mine. "It is for the sake of that being which I will be there at the turning of the path that I now exert all my strength, and it is in this sense that there is already a relation between my future being and my present being. But a nothingness has slipped into the heart of this relation" (31). So, "*I am the self which I will be, in the mode of not being it*" (32).

To be sure, Sartre would like to say that there is more than zero relation between present self and future self. To permit himself this minimal futurity of selfhood, he notes that "otherwise, I would not be interested in any one being more than another" (32). But how strong is this argument? Can he appeal to "interest", or selection, as the defining factor in having a future? It seems more consistent for Sartre to conclude that the future self is less about choosing what to become and more about having an abyss both in general, as well as in the particularly of its own self.

The important point is not whether I decide, but whether my future agrees to it or not. The decision I make now will be made by somebody who does not yet exist: A non-person on whom I already "depend" (32).

This does not mean that decisions are never made, but it means that they are virtual. It is like Bergson's theory that "pure" memory is never an actual image in the present (the latter is an image, not memory); pure memory is the possibility of calling up an image. Of course, I can picture the possible future, but the decision-image is not decisive. A genuine decision-content is only ever the possibility of deciding.

If this is a vertigo moment of anguish, throwing ourselves away into the abyss of time, we cannot counter it by deciding to be our true selves. We can only accept the nonidentity of the decision, to "transmute it into indecision" (Sartre *BN* 32). If I can allow not only that "indecision calls for decision", but also that indecision counts as decision, then, he says, "I put myself at a distance from the edge of the precipice and resume my way" (*BN* 32).

Sartre ends the diet entry with a good question: "Why record all this in such detail?" You might be asking me the same question. Ultimately, I just want to know how the temporal noema of a decision works. What is the takeaway from Sartre's lunch? Let me summarize bluntly. I decide now to do something in the future. When that future arrives, I will have to decide then whether to do it, so the decision I make now will be void. However as I know that already now, my current decision about the future cannot be about the real future. But it really is about the non-real future.

I am not sure that philosophy can help us much to keep to our decisions in practice. Can it at least explain why decisions are so hard? Is it helpful to have an existential-ontological explanation for why ecce homo weighs so heavily? Is it helpful for the dieter to know that he loses weight on some time lines but not this one? Or can indecision be a method for losing weight? Don't laugh, indecision is what made Sartre cut back. Hard and fast resolutions lead to disappointment, where irony might be able to live with fits and starts. I do not want to follow Sartre to the conclusion that the decisional future is a dead end. What I want is to put irony on a timetable, to give structure to the non-real future, to show how there can be multiple tracks on the time line of the real world as long as time is virtual, and to see if there might be a longer shelf life for decisions in multitrack.

CHAPTER TWO

Husserl: Decisions and temporal overlap

Not every angle on the topic of decisional futures is phenomenological, but my starting point is: What picture of the future is presupposed when we make open-ended decisions? As a first step, I will develop two lines of thought out of Husserl.

First, I find it useful to formulate the topic of decisional futures as a problem of the temporal noema of decisional acts of consciousness. What exactly Husserl means by "noema" in *Ideas 1*[1] is not always clear, so I will say a few things about it.

Second, and this will occupy most of my time on Husserl, I will develop some of Husserl's brilliant and odd examples in *Experience and Judgment*[2] of experiences in which different views of the world are juxtaposed at the same time. The conception of the future that I am aiming at involves just this sort of plurality: Acts of consciousness that envisage alternative, partly competing and partly converging, futures at the same time. Commentaries on Husserl have not focused on this aspect of his text, but in my view, it is central both to phenomenology as an approach to the world of experience, and to the particular topic of decisional futures. There is rarely anything better than Husserl at work.

Husserl himself does not always emphasize the multilayered character of phenomenological description. He does not always say that decisions project into plural futures. On the contrary, he sometimes speaks as though decision consists of taking a plurality of options and narrowing them down to just one. This idea is initially plausible, and no doubt some decisions have that form. Some interesting cognitive science, like Alain Berthoz's *Reason and Emotion: The Cognitive Neuroscience of Decision Making*, treats decision as the act of zooming in on one alternative instead of others.[3] On this model, decision is the act of suppressing alternatives.

However, in the final analysis, both in Husserl and in cognitive science, unchosen alternatives never disappear, but "stick around to try to win again" (Berthoz 64). In this second model, decision cannot rule the future, at best it can inhibit the return of certain options. Indeed, the "inhibition of return" may be a general feature of consciousness. Making a decision and not going back on it, may have a parallel in perception, where that "we rarely look back to where we have just looked when scanning a scene" (83).

Yet decision must do more than truncate the space of possibilities; it also has to root the chosen possibility in an actual present, in order to project the possibility into the real world future (268–70). Consequently, even if decision at first seems to be an attempt to reject plurality in favor of singularity, there are already at least two positions that any real world decision thinks in a single act of consciousness—one present and one future. Moreover, a decision is not simply two separate pictures—one present and one future—but is a projection from the first into the second via a dense sequence of intermediaries, a decision thinks many frames in a single act. Furthermore, the decision includes an ongoing comparison with alternatives, it thinks many series within a single act. Finally, as each step along the way feeds back onto the image of each other step, it thinks many series simultaneously. So even if a decision was paradigmatically the narrowing down of many possibilities into one, and not primarily a serial readjustment of each possibility in light of the others, it would still have a multilayered serial temporal referent.

My view of decision generally is that a decisional future is not just a forking path, but that when we choose one fork, we keep other forks in play, and can pick up the non-taken path part way through without going all the way back to the beginning. For example, if I decide to do something about global warming, I can choose between a recycling path, a political engagement path, a science education path, or something else. However, after pursuing one path for a month, or a decade, I might decide to pick up on one of the other paths as well, or instead. However by that point, I will have engaged in some actions that would also have been on the other path not taken, so I will not have to go back to its starting point. I will be able to engage at some midway point on the non-taken path, as if I had been following that other path as a shadow path all along. Indeed, I will have been following that path all along, but virtually rather than actually. In short, decisional futures are not just multiple forking paths, but convergence trees across forking paths in many different virtual patterns.

Before looking at the details of overlapping decisional futures, I will first argue that decisions have their own noemata, that is, their own objective referents, different from the objective referents of other experiential fields such as perception, desire, and prediction. In addition, I will argue that part of the noema of a decision is its temporal structure.

The noema of decisions

Husserl's *Experience and Judgment* is, as the title says, about the ways that perceptual experience generates logical judgment forms. For example, looking out at the world grounds first the "S is P" form, then the "S that is P is also Q" form, and then the "P and Q entail R" form, and so on. Much of the work to build this kind of transcendental logic involves analyzing the ways that perceptions transform from uncertainty to certainty.

Husserl's idea is that the perception that gives us certainty is only a limiting case (Husserl 271). The usual situation is that we anticipate more evidence than we get, or that we get more evidence than we can use, or that we are challenged by doubt, and in each case, we have to "resolve it by a decision" (272). Decision may "reinstate certainty," but certainty is not an immediate equivalence of thought and evidence; certainty is a "modalized" state, constructed as a possibility. Therefore, all of "the modalities of predicative judgment must be understood as modes of decision" (272).

This makes an interesting pair with what Husserl sometimes calls the principle of all principles, namely that all experiences presuppose the givenness of the real world, and that before anything is in doubt, the pre-predicative certainty of the world is given (Husserl *Ideas I* 44). That remains true enough, but it is nevertheless still the case that each and every predicative judgment needs a decision. What the pre-predicative horizon makes certain is just that there is enough potential evidence around us to make it worth trying to make decisions.

In the "Introduction" to *Experience and Judgment*, Husserl tries to locate a level of pre-predicative experience prior to any implicit judgments and cognitive achievements. The flux of perception looks like it fits the bill. However, even in the most immediate perceptions, whether in dreamy states or in buzzing confusion, we cannot help arresting the flow and attending to determinate shapes, if not full-blown objects. Even immediate perception includes achievements with judgment-like functions. Any more extreme pre-predicative level would not be able to include any focus, any objects, any shapes, or any themes. It is hard to see what that could be. Such a degree of uninterpreted givenness would be even less determinate than what Kant calls the sensuous manifold. Perhaps Michel Henri is right, and pre-predicative experience is simply the experience of there being experience, or the presence of mere presence, prior to the experience *of* something.[4] Husserl certainly says that there can be no identity in pre-predicative experience, as identity pertains to objects that we can return to under common judgments. Perhaps pre-predicative experience is meant only to cover simply the "wait for it …" function of experience. Or perhaps it is the experiential prime matter that perception is always already making decisions about.

However defined, there are both pre-predicative and predicative levels of "Decision." (a) At the pre-predicative level, decision covers passive syntheses of perception, where discrepant perceptions are resolved into coherence. (b) However, what Husserl calls "decision in the proper sense" is the "position-taking of the ego as an activity in the act of predicative judgement" (Husserl 272). Examples include "active decisions, convictions, allowing-oneself-to-be-convinced-by, taking-the-side-of, and so forth." There does not have to be perfect evidence or unanimity for a decision to take place, there only has to be a position-taking.

What is interesting to me is that "these activities also have their noematic correlates" (272). Husserl's vocabulary of noesis and noema means various things in various contexts, but in a nutshell, noesis and noema are the subjective and objective ways of describing the same experience. In some usages, noesis refers to the propositional attitude and noema refers to the object, so that the same noema, for example, a tree in the yard, can be apprehended in different noeses, for example, it can be perceived, or it can be loved, or its existence can be doubted. In other usages, noesis and noema are two correlated ways of describing the same concrete object, so that if the noema is a three-dimensional tree in space (objective description), the noesis is the perspective from which the tree is seen (subjective description).

There are difficulties in applying the vocabulary of noesis and noema to decisions, and these difficulties bring out some interesting points. The first problem is whether the decision is noetic or noematic. On one hand, there are situations where a person could have two subjective attitudes toward a single content: One could wish to play tennis, or one could decide to play tennis. Decision in that case is a noesis; the noema, the objective state of playing tennis, is neutral with respect to the noeses. However, on the other hand, if we describe situations in more detail, the decision seems to have a distinct noema. If a person decides to play tennis, he or she then has to think about buying a new racket, finding an opponent, doing some stretches, and so on. These intermediaries are not required for the attitude of wishing to play tennis. So in this case, which seems the norm once the examples are thickly described, decisions have their own noemata.

I will presume that deciding has a distinctive kind of noema. I have to admit that Husserl's position that decision and indecision are not "determining predicates" suggests rather that decision is about subjective attitude rather than objective reference. However, in my favor, Husserl sometimes says that "decisive position-taking" is not just one attitude among others, but brings a subject closer to the "essential and pregnant sense of judgement" (273). In addition, at least in one passage, Husserl refers to a "noema belonging peculiarly to the willing: 'the volition-meaning' " (233/199).

Much of Husserl's chapter on decision in *Experience and Judgment* describes the genesis of decision. In contrast with the "mere apprehension" of objects on the one hand, and with merely subjective "motives" on the other hand, decision points to a middle ground between object and subject. The genesis of decision begins in the flow of perception, where there are "continuous self-corrections," "constant partial cancellations," and cases of "letting one's glance wander over an object and fix on what is seen in a doubtful way" (276). The flow of perception itself thus contains modes of seeing "otherwise" than are immediately given. At this stage, corrections are passive; there is no need for "personal decision." As long are there is a "lack of contrary enticements," and "nothing speaks in the given moment" for an alternative, no explicit decision is needed. Whereas Heidegger thinks that decision is an ontological foundation of consciousness, for Husserl, we make decisions only when we have to. Nevertheless, we might at any time start to have to. For doubt can arise as soon as perception contains not-yet-present anticipation, and it always does.

All perceptions are thus in principle open to doubt. Doubt is the motivator of decision, yet after all, there is more room for decision when there is more perceptual evidence to work with. Husserl pictures this process as a move from the presumption of certainty to doubt, then to a decision that reestablishes provisional certainty. However the picture is more complicated, as the immediate presumption of certainty is already doubt-creating. Any judgment that exists within the flow of time is already "abandoned in its retentional reverberation." It is already becoming "indistinct" (279). We can at any time either "renew" a judgment about what we are seeing, or not. Sources of doubt, and hence the need for a decision, is built into the distance between a judgment and its fulfillment. Once the process moves back from doubt to certainty, we are no longer in the zone of decision. The in-between movement, the zone where we are free to draw our own conclusions, is the zone of decision. "Striving for decision" (282), made possible by the distance between the judgment and the experience that confirms it, is the epistemic source of cognitive freedom, and even of morality.

Decision's job is to resolve into not just any old result, but to bring experience to fulfillment and so to show what is true. All experience does this, but we especially call an experience a decision when it is "conscious of" doing so (283). To be conscious that a decision has to be made means "having before oneself the disjunction of truth and falsehood" (284). The disjunction Husserl has in mind here is not a disjunctive set of alternative interpretations, but the disjunction of the truth of the situation on the one hand and the set of all false expressions on the other. Disjunction thus appeals to the division between the real and the possible. However, decision operates only when all positions still look like they might be the real one. Decision takes as its object a plural, disjunctive, no-man's-land, not merely imaginary (as the point is to decide which is real) and not yet affirmative. The final goal in making a decision may be to verify a judgment objectively, but decision takes as its material a set of intentional objects all of whom are still "separate from the actual state of affairs" (284). Getting *away* from actuality, preserving both the real and the possible as equals in a common virtual state while the decision is being made, is the "critical attitude" (285).

Husserl works out four cases: Where the doubt calling for decision looks like disagreement, conjecture, conviction, or presumption.

Disagreement: The first case shows an ego in a state of "cleavage," or "disagreement with itself," "inclined to believe now this, now that" (303). It does not sound right when Husserl describes uncertainty as alternating from one certainty to another. However, uncertainty does represent a state of divided commitment or divided compliance, and in that sense, doubt does look like alternating futures. Perhaps this is why Husserl says that uncertainty is intolerable, and produces a striving for certainty. If uncertainty were simply the cognition of degrees of probability, it would be a kind of knowledge in its own right, and hardly intolerable. However, if doubt is the contradictory inclination toward two different futures, then it is unstable.

"Inclination" toward one possibility rather than another is not just subjective, but represents the "reflective pull" or "the weight of the possibilities themselves." "The possibilities attract me in their being" (303). In this way, both noesis and noema are parts of the "personal decision." The noesis is the "impulse to act," "feeling oneself drawn in," and "striving"; the noema is the "position taking" that "espouses one of the two sides." It is the noematic force that gets a perceiver to follow a possibility.

In any case, the call for decision operates midway between uncertainty and certainty. When Husserl speaks of alternating compliance, it almost sounds as though we make resolute decisions but alternate them, that decision is not just a reaction to alternative hypotheses, but actually takes up alternating resolves one after the other. What would it actually mean to be decisive in that way? Does the paradigm decision take the form of electing Democrats and Republicans, or Anglophone and Francophone prime ministers, in alternate cycles? Is alternating polygamy more decisive than lifelong monogamy?

Conjecture: Sometimes Husserl suggests that doubt persists only until one possibility has a greater than even likelihood of being true (304). However, obviously doubt can persist even if one possibility is likelier than another. Husserl talks as if the very imbalance of strength in probabilities is enough to "motivate a decision," allowing us to "take one side" even while we admit the other side has some weight. The picture of evolving decisiveness is a good description of shifting decisions in the process of advancing research. On the downside, such descriptions say nothing about creating new theses to evaluate, or about experimenting with fictive futures.

Conviction: In cases of "impure affirmation," we make decisions in spite of having only insufficient grounds, being unwilling to be swayed by the negative evidence. Conviction in this sense is "decision with a bad logical conscience" (305).

Presumption: "Presumptive" decision-making is close to not making a decision at all, a hedge retaining the "open possibilities" even after the decision is taken (306). Husserl presents this last case almost as an exception to the nature of decision, but I think the futural openendedness of presumptive decision-making is the normal situation of the lifeworld.

What ties these four cases together is "questioning as the striving for a decision by judgement" (307). Noetically, questioning is an "impulse" occasioned by "discomfort" (308). It may not be obvious why unanimity must be the default comfort zone, but it seems fair to say that a question calls for something, and that in this way "arises a striving."

All modalities of decision arise from having a plurality of conflicting intuitions (A, B, *and* C), each one of which appears as a "problematic possibility," then asking how to settle them (A, B, *or* C). In some cases, none of the given possibilities will be satisfactory, or some will not be explicit, and the decision will send us out to look for further possibilities. As the answer fulfils the question, we might say that the answer is the direct noema of the question, or even that the answer to a question is the noema of the question's noema. The downside of this analysis is that as many possible answers are anticipated in the question, it assigns to a single question many incompatable noemata. Another downside is that "I don't know" would not really count as an answer to a question. It may be meaningful to say "I don't know," or "I am undecided," or "I don't know for sure, but probably X," but these would be different from the strived-for noema.

Another consequence of treating the decision as the noema of the question is that the noema will undergo a genesis in the course of decision-making. The content of a fully made decision, with no loose ends, will be a singular state of affairs; whereas a still undecided decision procedure will continue to contemplate disjunctive content-alternatives. In most decision procedures, we can always keep asking "Is it really so?," and look for more justifications than we have found so far (311). This is what it means to say that "decisions are based on reasons" (311). It is not just that a certain decision fits best with evidence. It is also that there is a genesis (a history with its future) of refinements and forced confirmations that ideally leads to the vanishing point of alternatives. In short, a judgment can be instantaneous, but a decision is genetic. Precisely because it is not finished engaging with the world, decision is more objective than judgment. Paradoxically, the "impure certainty" of "conviction," which leaves a taste of bad conscience about sticking with one's own judgment, is a better route to reason than the feeling of confidence with one's judgments.

The temporal noema of decisions

Sometimes, when one refers to an event, one does so without temporal reference, as when one refers to "getting up in the morning." In that case, the same objective reference is repeated on different mornings. The time does not enter into the noema, it only pertains to the occasion on which it is spoken of, which is more relevant to the noesis. However, if one refers to "getting up in the morning of New Year's Day 2017," the time referent is built into the noema. If one makes a New Year's resolution to exercise at least three days a week throughout 2017, the temporal referent is a kind of disjunction, but its meaning is clear and definite. If one's New Year's resolution is to be more honest throughout 2017 even when it is difficult, the year referred to is clear, but the days, if any, on which one's honesty will be tested are unknown and indefinite. However, there is just as much a temporal referent in this meaning content as there is in the more definite cases. If one's resolution has the form "I will either have a child or write a book," the temporal noema is complicated both by the disjunction, and by the fact that each disjunct implies many steps of different duration, and also by the fact that both of the disjuncts may well alternate or overlap during the year. All decisions have temporal reference of one sort or another, and the more complex the decision, the more its meaning hangs on the meaning of its temporal reference.

Furthermore, there must be a direct connection between the possible future continuations of a decision and the present time at which the decision is made. Otherwise, it will not be the present decision that is carried out in that future. Of course, the future continuations cannot be given in experience with the same intuitive presence or certainty as the current decision. After all, the future has not happened yet, and is not there to be intuited. The future, obviously, will reveal new aspects of the world not available to current experience. So the futural content has to connect to the present, but in a virtual and non-real way. The current content is given in real intuition, and the future content, which is a part of the same essence, is given in a different, non-adequate, way.[5]

In this chapter, I will focus on Husserlian resources for describing ambiguities, overlaps, and disjunctive relations across layers of the temporal reference for decisions and other sorts of experience.

Overlapping decisional futures

This is what I use Husserl for: A theory of overlapping futural time lines—some actual and some virtual. Throughout *Experience and Judgement*, Husserl develops five cases of overlapping time line phenomena that split across actual and virtual time. In each case, there is a slightly different relation between fixed and unfixed time. To some of Husserl's descriptions, I will make what seem at first to be small, friendly amendments, but which in the long run have major implications. Husserl's five cases of time-overlap consist of:

1 Mistaken identity, where two perceptions occupy the same space time. This is the case Husserl calls "overlap" (*Überschiebung*).

2 Fictive imagination, where two versions of a situation are articulated without there being any fixed temporal relation either across the different narrative versions of the story, or to any actual perceived objects. Where a fiction stands in no temporal relation to something in actual time, but

nevertheless has some temporal structure to its narrative, Husserl says it occupies "quasi-time" (*Quasi-Zeit*).

3   Memory–Perception overlap, where a subject remembers an object, say a table in front of her, at the same time as she perceives the same table actually in front of her. Where memory and perception line up and coincide in space and time, Husserl calls it "juxtaposition." I will add the term "juxtaduration" to describe its consequences.

4   The cognition of irreal objects, such as mathematical objects, which have the same content when experienced at different times. Husserl calls this "supertemporality."

5   Retained counter-motives. When a decision has to be made, it is because there are different motives tending toward different decisions. Once a decision is made, all those motives for the non-chosen decision possibilities move to the background. For various reasons, those motives can return and force a different decision to be made later on. On my interpretation of Husserl's discussions, this means that decision paths not taken continue to be lived virtually, and can at a later date return to actual force.

Overlap

Overlap arises in an experience where we are inclined to make a predicative judgment, but vacillate before we decide which judgment to affirm. Sometimes, Husserl says, our perceptions are interrupted and disappointed, and we simply cancel the activity of judgment for the time being. However, in the case he studies here, "in place of simple cancellation, we have a mere becoming-doubtful (*Zweifelhaftwerden*)" (91).

Husserl's case of uncertain identity is the old favorite: Man or Mannequin?

> Perhaps we see a figure standing in a store window, something which at first we take to be a real man, perhaps an employee working there. With closer observation, the doubt can be resolved in favor of one side or the other, but there can also be a period of hesitation during which there is doubt whether it is a

man or a mannequin. In this way, two perceptual apprehensions overlap. (Husserl 92/99)

The term for "overlap" or "overlay" is *Überschiebung*. What is interesting is that Husserl does not say that we see one thing (say, the man) and imagine or speculate about the possibility of the other (the mannequin), or vice versa. He says that we stand in front of the store and actually have two different perceptions at the same time. Moreover this seems right. The recognition that we might have mistaken the identity of the thing behind the window is not merely "a decisive (*entscheidende*) disappointment," where a new judgment replaces an old perception.

> Rather, the full concrete content in the actual appearance now obtains all at once a second content, which slips over it: the visual appearance, the spatial form imbued with color, was until now provided with a halo of anticipatory intentions which gave the sense "human body" and in general, "man"; now there is superimposed on it (*schiebt darüber*) the sense "clothed mannequin." Nothing has changed regarding what is really seen; indeed, there is even more in common: commonly perceived on both sides are clothing, hair, and the like, but, on the one hand, flesh and blood and, on the other, probably painted wood. One and the same complex of sense data is the common foundation of two apprehensions superimposed on each other (*übereinander gelagenten*). (Husserl 92/99–100)

Of course, it depends on what is meant by "apprehensions." When Husserl says there are two of them superimposed, he does not mean that there are two sets of "sense data" (*Empfindungsdata*). There is only one set of sensory material, and that is why it is so odd to have two apprehensions of the same data. However, it is not that there is one perceptual experience and two judgments that could be made about them. The point is that we do not know which judgment to make. We hesitate before making a judgment. If we think of apprehension as a perceptual experience directed toward perceivable objects, then we can say that we have two apprehensions, two perceptual acts of consciousness at the same time, in the same position in the world. We see the world twice at the same time and place. Each is "motivated," but they "stand in mutual conflict." In doubt, there remains an "undecided conflict" (*unentschiedender Streit*) (92/100).

> We therefore have, as it were, a bifurcation (*Auseinandergehen*) of the original normal perception, which in unanimity constituted only one sense, into a double perception (*Doppelwahrnehmung*). They are two perceptions, interpenetrating each other by virtue of the content of their common core. And yet not really two, for their conflict also implies a certain reciprocal displacement. If the one apprehension takes possession of the common intuitive core, if it is actualized, then we see, for example, a man. But the second apprehension, oriented toward the mannequin, does not become nothing; it is suppressed, forced into the background deprived of its power. Then, perhaps, the apprehension 'mannequin' suddenly obtrudes; we see a mannequin, and it is the apprehension 'man' which is put out of action, suppressed. (Husserl 92–93/100–101)

It is not just that the two perceptions alternate in turn. They each persist even while the other appears as the truth. As soon as there is once doubt, resolution of the doubt has only a temporary value. The option decided against "does not become nothing," and can return. We will see this scenario again near the end of *Experience and Judgment* in a different context. It is the status of the suppressed half of the double perception that preserves the other, second, lifeworld in virtual status. Indeed, "neither one [neither the perception of the man nor the perception of the mannequin] is there in the same way that the perception of the man was there before the onset of the doubt" (93/101). Once there is doubt, every perception, including the most obviously true ones, persist as much at the virtual as the real level.

In short, the indecision state exists in the unharmonious bifurcation of a single perception into multiple virtual realities. The indecision state belongs in pre-predicative experience just because it is preresolving, that is, it does not resolve into one object or another. It belongs in the non-present of the experience—not exactly in its future, as its referent is no more a determinate future than it is a determinate past or present—but in the multiple time lines that continue forwards from it. Moreover, it belongs to the reality of the lived world, not just in subjective imagination.

This problem of double, in principle multiple, perceptions is inherent in pre-predicative experience, where the content is not itself a predicative judgment but the perceptual attitudes to the world that precede definite predications. Husserl calls this "multi-sense (*mehrsinniges*; Churchill and Ameriks translate it as "equivocal") consciousness." In addition, he adds that this "fact of being split in two with its apperceptive overlapping (*Zweispältigwerden*, literally becoming-split-in-two) is continued in retentional consciousness" (94/102). The split, as we will emphasize later, is preserved in memory, in our experiential habits, and in the motives and tendencies toward further interpretation of the world that we carry with us over time. Or if we want to put this in terms of the sort of ego who has such experiences, we can say that "the ego vacillates (*schwankt*)" (95/103).

Pre-predicative experience might therefore be called the zone of indecision, and pre-predicative indecision could be thought of as the ground for those later modalities of judgment of which definite judgment is just one mode.

Quasi-time

In the second and third cases of superimposed apprehensions, actual and virtual variants coexist in one experience of the world. These arise because of the temporal structure of passive apprehension. Such cases are at first presented as exceptions to the norm. However, Husserl's scenarios in principle cover a great many sub-cases, and in the end, they seem at least as common as the apparently normal cases.

In describing the normal situation of perception, Husserl says that a "sensuously constituted temporal series is unique in every respect." That is, "all *that appears originally*, even if it appears in conflict, *has its determinate temporal position* (*bestimmte Zeitstelle*), that is, it has not only a phenomenal time, that is, a time given in intentional objectivity as such, but also *its fixed position (seine feste Stelle) in the one objective time*" (164/190). There are two ways to take this principle: That each event occurs at one fixed time only, and that each fixed time includes only one set of events. Both will be at issue.

Husserl emphasizes that the issue is not whether we can experience an object at many subjective times. Obviously, we can. Nor is the question whether there must be a unity of subjective time. Husserl figures there must be, but that is not the point here. The point here is that when we experience an object, we have to think of that object as fixed in time. In other words, the point here concerns noematic, rather than noetic, temporal fixture. The points we raise here about fixed and unfixed time relative to perception will later apply equally to decision.

Husserl's claim about fixed times does not imply that there are instants of time, which would be both implausible and not very Husserlian. However, it does imply fixed durations. This assumption of fixed durations would not be easy to articulate and defend, but I do not want to worry about this now. Even if we would better say that time segments are blurred at their starting and end points, we might still somehow be able to say that these are in some way "fixed," coordinated both in a subjective stream and in terms of the relations among objects presented simultaneously or successively in it.

Husserl's motive for this view is broad. For one thing, he appeals to "the Kantian thesis," essential, Husserl thinks, for any rigorous transcendental analysis, that "all given times become part of *one time*" (164/190). For another thing, Husserl anticipates that the unity of time will be required for empathy (165/191), presumably as a condition for different egos to share a lifeworld over time. As we consider exceptions for the fixity of intentional objects in time, we will effectively challenge not only the Kantian one-time thesis but also the ethics of empathy.

The first exception that Husserl gives to the thesis that every intended object has a "fixed position in one time," is the case of fictional objects: *Fikte*, "lived experiences of imagination, which are directed toward fictions" (167/195). There is an obvious sense in which fictions are not fixed in time. As they are not descriptions of actual objects and events in the actual temporal world, they cannot be tied to a fixed time. Even if, inside the fiction, it is said that an event occurs at a fixed time, that time is designated in the fictional world, and is still not fixed to the actual time. A story written in 1948 may say "now," or may say "1984," but that does not fix the fictional event either to the real 1948 or to the real 1984, or to the real 2017 when it is being read. In addition, imaginary objects are even less datable:

> The centaur which I now imagine, and a hippopotamus which I have previously imagined, and, in addition, the table I am perceiving even now have no connection among themselves, i.e., they have *no temporal position in relation to one another (keine Zeitlage zu einander)* ... the centaur is neither earlier nor later than the hippopotamus. (Husserl 168/195–6)

To be sure, there can be temporal relations within the fiction—the hippo can age within the fiction. Moreover, there are always temporal relations in the subject who invents the fiction and in the subject who reads it, and between writer and reader, and in the history of the book. "The object of imagination is present to consciousness as temporal and temporally determined, enduring in time: but [*Here is the interesting point*] its time is a *quasi-time*" (168/196).

Quasi-Zeit. "It is a temporal object, it has its time. And yet it is not in time" (168/196). Again on the next page: "Time is certainly represented in imagination, and even represented intuitively, but it is *a time without actual, strict localization of position (Örtlichkeit der Lage)*—it is, precisely, a *quasi-time*" (169/197).

This idea of quasi-time is developed to describe fictions. However, fictionality is not what defines quasi-time, it is only a pertinent example. What defines quasi-time is that it covers temporal orderings that are not fixed to segments of the time line of the actual world. In addition, this definition will fit other phenomena broader than fiction, including decision contents and indecision contents. Because it is so broad, Husserl develops many variations of quasi-time over the next several pages of the text. Husserl's analysis of imaginary objects describes "quasi-perception" and "quasi-recollection" (170/198), "quasi-world" and "quasi-fulfilment" (171/199), "quasi-actuality" (172/200), "quasi-intention" (172/201), "quasi-individual" and "quasi-identity" (174/202). For Husserl, the range of temporal relations across imagination objects is a wide open field of enquiry: "every act of imagination, being divorced from all [temporal] connection, has its own *imagination-time*, and there are many such ... infinitely many" (170/198). To put it in the strongest way: "Everything is now carried over to the quasi (*in das Quasi*)" (172/201).[6]

Although Husserl treats fictional objects in quasi-time, he does not want the properties of quasi-time (an object's detachment from the actual fixed time order and the openness of the fictional object to be relocated elsewhere in the time line) to extend to actual objects. Quasi-time applies only to a "possible world" (*'möglichen Welt'* [Husserl's scare quotes]) (173/202). Husserl contrasts this with the surprisingly blunt principle: "In the actual world (*Wirklichen Welt* [no scare quotes this time]), nothing remains open; it is what it is" (202/173). It is surprising that Husserl would separate in this way an actual intentional object from its possible continuations, as these possible syntheses of identification and levels of fulfilment define and circumscribe all meaning intentions. We can appreciate Husserl's motive for fixing objects in time. In some sense, it is obviously correct to distinguish actual objects in time from fictional objects that can be imagined to exist at any time whatsoever. However, the hardness of Husserl's distinction is about to lead to a claim that, it seems to me, is too strong. Moreover, this will be the point where I will offer a friendly amendment to Husserl's description that will in the long run lead to a significant divergence.

Husserl's example is the Hansel and Gretel fairy tale:

> How are the singularizations of temporal points, temporal durations, etc., related to one another within different imaginary worlds? We can speak here of the likeness and similarity of the components of such worlds but never of their *identity*, which would have absolutely no sense; hence, no connections of incompatibility can occur, for these would indeed presuppose such identity. It makes no sense, e.g., to ask whether the Gretel of one fairy tale and the Gretel of another are the same Gretel. (Husserl 173/201–2)

This is similar to David Lewis' view that things in different possible worlds cannot be identical (see Chapter 9). However, it cannot quite be right. Of course, one variation on the Gretel tale is not refuted by another; they are not competing to be the one true story. Moreover it is true, as Husserl says, that the time sequence of events "within the same tale" can be determined straightforwardly (in every version, the witch captures Gretel before she pushes the witch into the oven, not after), whereas events across variations are not ordered in the same way (the Gretel who pushes the witch into the oven in one tale does not do so either before or after or at the same time as the Gretel in a different version pushes the witch into the oven). On the other hand, when the recent film *Hansel and Gretel: Witch Hunters* (Tommy Wirkola, 2013) puts a makeshift machine gun into Gretel's hands, it is clearly that same Gretel from the fairy tale that is being given a new twist. So there is a synthesis of identification at work across versions. Moreover, while this film is semi-updated, it retains strong contextual reference to the premodern world of more-or-less medieval Germanic culture. Viewers certainly know when it is time for Gretel to push the witch into the oven in the movie, just because we know when it was time for her to do so in the old tales. To extend the example field, fictionalized stories of real people, and documentary as well as mockumentary fictional characters, presuppose a degree of temporal non-fixity in nonfictional objects, as well as a degree of temporal fixity in fictional objects.

Husserl does at times qualify the distinction between perceptions fixed in time and imaginations unfixed in time, "in that imagination involves more than a merely indifferent parallel to actual experience" (Husserl 174/203). Without that qualification, Husserl's division does not work. It is fortunate for the problem of decision that the distinction does not work. For the key to decision is that the various possibilities we imagine before deciding are non-actual during deliberation, and yet are all fixed to the same temporal moment at which a decision will direct future actions in one way or another. In decision, imagination objectivities are hinged to real time. I noted above that there are several senses in which objects are fixed to time: The sense in which each event occurs at just one time, the sense in which only one set of events occurs at a given time, and the sense in which all events are ordered continuously in just one sequence. Different sorts of objective noemata may be fixed in one or more of these senses but not in all senses. Imagination objects may be fixed to actual time but may be experienced at more than one time. Decision objects may be fixed to time, but the branching time lines engaged by a decision may not be unified by a single temporal sequence. Or the time in which decisions are made may be considered a unity (unified by the possible convergences across decision paths), but time-unity may not be the same as univocal time-order. In brief, there might be time-fixity without time-unity, or time-unity without time-order. We will need these variations in order to analyze the future of decisions.

Juxtaduration

The third case of superimposed apprehension, and the second case that Husserl calls "unconnected with regard to their temporality" (180/211), involves the superimposition of a memory onto a perception.

> If we place the remembered table beside (*neben*) this perceived table, then we have a space with a spatial plenitude and, giving itself in it, a vivid second table and a time in which this juxtaposition of both tables appears for a while (*eine Weile*). Here it does not matter that the remembered table in itself "belongs" to another objective time than the perceived table. We have a unity of "image", and this is the image of a present, of a duration with a coexistence (*Koexistenz*) to which pertains a spatial unity. Thus we can spatially "bring together" objects belonging to different fields of presence (*Präzensfelder*) if they are physical objects, "juxtapose" them temporally (*zeitlich nebeneinanderstellen*) or bring them back together temporally (*aneinanderrücken*) ... We can then say: we bring objects which belong to different fields of presence together by transposing them to *one temporal field* ... into a unity of simultaneous duration ... contemporaneity as enduring side by side (*Gleichzeitigkeit als nebeneinander Dauernde*) or as appearing one after another in this space and remaining there for a while (*vorweilende*). (Husserl 181/212–3)

This is a nice example to use in a public talk on phenomenology. The speaker and the listeners are all looking at the same lectern for an hour or so, and everyone can remember small differences—when the water bottle was moved, or when the speaker walked around it. So when the speaker asks the audience members to experience the lectern from a while ago at the same time as they are seeing it now, they can try. Some members of the audience report that they can do that, others report an inability to have both experiences at once, and still others report that they do not know what is being asked of them. Unlike the case of overlapping perceptions of man and mannequin, the idea here is not to have two experiences of the same noetic type (two perceptions, only one of which can be objectively true in the final reckoning) at the same time, but to have two different sorts of experiences (one memory and one perception, both of which are objectively true) at the same time, both targeting the same intentional object, but each targeting the object at different moments of the object's duration in time. It may be that some people cannot do it. However in principle, there does not seem to be any contradiction in having two such experiences at the same time, so there is no reason to doubt those who say they can. Where Husserl called the case of double perception "superimposition", he calls this case of time doubling "juxtaposition."

Temporal juxtaposition of experiences of an object, *zeitlich nebeneinanderstellen*, gets its name from the idea that the two table objectivities, though separated by their temporal reference, exist cognitively "beside" (*neben*) one another in a single moment of consciousness. To be consistent with the translation of the prefix, we should translate *nebeneinander Dauernde* not as "enduring side by side" but as "juxtaduration." Moreover as the *stellen* in *nebeneinanderstellen* is the same term that Husserl uses in his definition of decision per se as "position-taking" (*Stellungnahme*), we should understand the situation where we experience the same object in different ways according to its different temporal states not just as "juxtaposing," but as "juxtaposition-taking" or even "juxtadeciding" or "juxtacision." Furthermore, if we put the emphasis not just on the *neben* in *nebeneinander*, but also on the *ander* in the same word, we could also understand juxtaposition-taking as "other-position-taking," that is, as taking whatever position is directly present on an object as well as another position also.

The category of juxtatemporality covers cases where an object can be identified according to two quasi-temporalities that coexist not as fictions, but as actual times. Because it allows two actual times to be present simultaneously without conflict, it allows for temporal unity without unique subjective temporal ordering.

Supertemporality

The fourth case of superimposed perception arises some eighty pages later in the text, and covers an extremely broad range of phenomena: "irreal objectivities." The example range begins with mathematical and logical objects. "Irreal objectivities make their spatiotemporal appearance in the world, but they can appear simultaneously in many spatiotemporal positions and yet be numerically identical as the same" (260/213). Different people can think of them at different times, but their truth is always the same, and in that sense, their truth subsists simultaneously at the different times of their instantiation in peoples' minds.

Once this familiar argument is used to cover mathematical cases, a much larger range of cases then suits the principle. Any "proposition," from "roses are red," to "Nixon lied," to the infinite propositions that have never yet been uttered at all, can be articulated at many different times. This is a principle central to Husserl's antipsychologism, first stated in the "Prolegomena" of his *Logical Investigations*. In general, Husserl says, using the phrase he used to define fictions, "the *proposition itself* has no binding temporal position (*Bindende Zeitlage*), no duration in time" (259/311).

Or to say it better, a logical proposition is not so much timeless as privileged in time by its generalized temporal availability:

> *The timelessness of objectivities of the understanding*, their being "everywhere and nowhere", *proves*, therefore, *to be a privileged form of temporality* ... That is, a supertemporal unity pervades the temporal multiplicity within which it is situated: this *supertemporality (Überzeitlichkeit) implies omnitemporality.* (Husserl 261/313)

A few pages later, Husserl gives a more precise definition of "supertemporality." A "free ideality" is "repeatable" (*wiederholbar*) at different times. Fictional objectivities were earlier defined as objects unattached to any time; ideal objectivities are now defined as objects attached to all times. In addition, perhaps not surprisingly, objects very much like fictional objectivities, now designated as "cultural objectivities," are put into the category of free idealities. Husserl's examples include Goethe's *Faust* and Raphael's *Madonna* (266/319–20). Goethe's *Faust* is a reprintable text "found in any number of real books." And although Raphael's painting is unrepeatable (Husserl is probably right not to be concerned that Raphael made many different paintings on the Madonna theme), its ideal image is in principle as repeatable as a text. Indeed, the scope of ideal objects, which started with mathematical objects, extends beyond fiction as well, and includes such objects as civil constitutions (266/320), presumably in the sense that a constitution is a precedent appealed to on all relevant occasions.

The scope of objects without a unique fixed position in actual time is by this point so broad that it is no longer possible to consider them as exceptions to the normal relation between objects and time. Relaxed temporal fixity means precisely that (a) a given object can in different ways be both present and not present at a given time, and (b) a given time can in different ways contain both an actual object and that object's non-actual variations.

In one sense, super-time is closer to quasi-time than it is to juxtatime, in that it unfixes the temporal location of a given object so that it can be inserted at any point in the time line. However in another sense, super-time keeps no-longer and not-yet objects active in the time line, and so generalizes the memory-perception doubling, suggesting that juxtatime is the more fundamental form of time.

Decisional oscillations and retained countermotives

We now come to the temporal structure of decisional contents. Husserl says that "the expression 'decision' is ambiguous." "Decision in the proper (*eigentliche*) sense" (272/327) brings all doubt to an end. However, decision-making in the broad sense covers the whole process starting with naïve certainty, followed by oscillating judgments, then provisionally taking a side, going through a period of stable conviction, then possibly followed by subsequent stages of renewed doubt. The more we consider Husserl's analysis of the latter, the narrower the scope of decision in the "proper" sense will appear.

In order for there to be a decision, there first has to be some question to solve, or a set of alternatives to choose among. An immediate judgment is not a decision. Or to say it better, so-called immediate judgments are short-circuits in the decisional process: "All modalities of judgement must, on principle, be conceived as modes of position-taking, as modes of decision" (289/348). Even the simplest judgments presuppose decision, and decision presupposes a prior state of indecision.

There are two necessary conditions for being in a state of indecision. First, "there must be countermotives (*Gegenmotive*) ...; disjunctive possibilities in reciprocal tension" (274/329). When there is a "lack of contrary enticement" (*Gegenanmutungen*), there is no decision to make (277/333). "The decision *for* one of the possibilities has juxtaposed to it, as a correlate, the decision *against* the correlative possibility—if not actually, then at least potentially" (290/349).

Second, the ego must "strive" (288/348) for a resolution. The vocabulary of "striving" has a bad name, on account of Fichte's notion of striving and Hegel's sharp criticism of it (in *The Difference Between Fichte's and Schelling's Systems of Philosophy*[7]). For Fichte, the self never apprehends objects as such, it only apprehends the way it is limited by a not-self that it projects outside itself. For Hegel, this would condemn the self to a hopelessly unfulfillable desire for confirmation. Politically, Hegel says, a self capable only of striving would desire the state to keep it under constant police surveillance, to confirm endlessly that it is still recognized in a world kept secure. However, such an insecure self would effectively know that its sense of security is a false one, and so would devolve into a kind of paranoia, suspecting that all signs of social reality are counterfeit. Phenomenologically speaking, "striving" for reality falls short of intentional objectivity. Moreover, the first principle of Husserl's phenomenology is surely not that we strive to make objects present; it is that objects are given as present. In short, striving is not a great way to describe how we aim at experiencing those objects given without complete certainty. Nevertheless, it is fair to say with Husserl that in a situation where different interpretations are possible, the alternatives are not just different descriptions, they are different claims to the subject's assent, competing to be taken as true. The presence of alternative countermotives will then be expressed in a series of "alternating compliance" (303/366), in "oscillations of certainty" (277/333). We might resort to saying that we "strive" to resolve doubt.

Husserl lists three sources of doubt, that is, three situations in which different possible judgments will compete for truth. The first two are straightforward and Husserl cites them just briefly.

First, doubt "will everywhere be the case where the act of predicative judgment does not take place in complete originality, on the basis of the completely original self-giving of the judicative substrates" (275/330). This is Husserl's way of referring to the situation where there is incomplete perceptual evidence (i.e., incomplete original self-givenness, or incomplete intuitive fulfilment, etc.) for the object of judgment. Husserl does not emphasize this here, but there are basically no objects which are ever given in complete perceptual evidence. Three-dimensional physical objects, for example, are given perceptually only one perspective at a time. As every empirical judgment is contingent, or as Husserl puts it, "autonomous," it can only be "provisional, as an indeterminate truth" (288/347). In short, this criterion for determining when there will be doubt in judgment covers virtually every judgment that can ever be made.

The second source of doubt arises whenever a judgment is expressed in a way such that more "precision" could be possible (276/371). Again, that covers just about any judgment whatsoever.

The third source of doubt is, if anything, even more pervasive. A decision begins with indecision over competing judgments and their countermotives. Decision puts an end to indecision. However, the very fact that there were once alternatives implies that there will always be alternatives. Moreover this entails that every possible decision is subject to a return to doubt. Husserl analyzes this situation in three steps. First, certainty recedes. Second, decisions are devalued. Third, rejected decisions live on.

The first step in the return to indecision, or better, in the overlap of decision and indecision, and therefore in the preservation of alternative futures within the same decision, or in other words, "the first modification [of decision], is that of *retention*" (279/336). As it is obvious that every content of consciousness is retained in what Husserl calls primary memory, even if it is forgotten in explicitly conscious, or secondary memory, it follows that once consciousness has considered motives for alternative decisions, all those motives are retained somewhere in implicit consciousness, even after a person has chosen in favor of one of those decisions over the others. On the one hand, Husserl says that once a decision has been made, one of the alternatives may be "abandoned in its retentional reverberation. It then sinks ever further into the background and at the same time becomes ever more indistinct; the degree of its prominence gradually lessens until it finally disappears from the field of immediate consciousness, is 'forgotten'" (279/336). However, on the other hand, Husserl continues in the same passage:

> It is henceforth incorporated into the passive background, into the 'unconscious', which is not a dead nothingness (*kein totes Nichts*), but a limiting mode of consciousness and accordingly

can affect us anew like another passivity in the form of whims, free-floating ideas, and so on. (Husserl 279/336)

If past motives, as free-floating ideas or whims, functioned like free idealities, that is, like propositions entertained but no longer believed, then unmade decisions would be remembered but would in no sense posthumously be remade. However Husserl's line of reasoning goes further.

The second step in the return to indecision arises because the earlier grounds for making a different decision did in fact have some rational force. That rational force does not go away just because different evidence outweighed it. Indeed, the longer we live, the more decisions we have made, the greater our accumulation of reasons for making a great variety of different decisions.

> There is a ground composed of all the judgments already passed ... We have a progressive taking-cognizance-of, an adaptation of the knowledge previously acquired to what arises anew ... At the same time, the possibility exists in each case that the knowledge already effected as valid and made part of one's habitual possession, instead of becoming united with the new—enriching it, completing it, and determining it more precisely—is, on the contrary, modalized in a *negation* (is canceled), or again is modalized in *another* way: is nullified disjunctively in doubt, in mere conjecture, etc. From this springs *the striving for decision* and the necessity of a criticism of the judgments already passed. (Husserl 281–2/283)

It is another obvious fact that even when there is plenty of evidence present for a given decision at the time the decision is made, by the time we reflect on that decision later on, that evidence is no longer present or current. It is thus inevitable that to any reasonable person, decisions will look more precarious as time goes by. It is not only inevitable, but right, that we engage in "criticism of judgments already passed." Indeed, if we ask how soon after a decision is made we ought to start criticizing it, the answer is right away. "As soon as [predicate judgments] have been produced," they become "autonomous" relative to the evidence they were based on, and so become open to a new assessment (285/343). At the very moment the decision is fixed, it becomes unfixed again, much as Sartre said.

Husserl describes the situation in a powerful and touching way: As "the growing future of devaluation of the results of the judicative position-taking already obtained" (282/340). (This is not the only occasion when Husserl uses the economic metaphor of capital depreciation and hedge funds in cognition.) This is the destiny of each decision, and of a person's entire history of decision-making: A "growing future of devaluation!" But how could it be otherwise?

> What speaks in favor of this possibility is precisely the general experience of the frequent reversal taking place in judgment, but in the given case nothing in addition to this; in the actual context itself, *everything* speaks in favor of our certainty: it is and remains certainty, but, to be sure, a certainty which has, in addition, a counterpossibility; it is a certainty, therefore, which has lost its purity. (Husserl 282–3/340)

Even when there are no actual grounds for doubting a decision already made, the simple passage from one content of consciousness to another over time generates the momentum for doubt.

The more certainty, the more doubt. Decision passes through criticism. "And this is the normal case (*Normalfall*)" (284/342).

This leads to the third and final step in the return to indecision. Not only does the certainty of past decisions lose its purity in general. The rejected decisions gain power. What happens to a possible decision when it is decided against? Where do abandoned virtualities go? To answer this, we have to acknowledge that decisions are more complex than Yes/No alternatives.[8] On the one hand:

> The nullified apprehension with its nullified intentional tendencies, above all with its still living, still dynamic, but canceled expectations, can no longer be carried out. (Husserl 290/292)

Earlier, Husserl had said that the canceled judgments were "not nothing"; now he says they are still living (*lebendig*). His term for "nullified" here is *aufgehobenen*. Husserl does not likely have Hegel's overcoming/survival specifically in mind, but he uses many other terms for "nullification" in other passages, so his choice of this term here is deliberate. The point is that while the rejected decision option is definitely nullified, and can no longer be carried out—after all, it is decided against—it remains a force in future reasoning. On the one hand, Husserl is careful to say officially that only the judgment that the decision declares valid "has the character of being valid *henceforth* (*hinfort*), of continuing to be valid later on; this means a validity within an open temporal horizon of a conscious ego" (290/350). However, on the other hand, he does not let the question of the futural continuation of the rejected decision drop. On the next page, he raises it again:

> What, then, happens to the opposite apprehension which has come to nothing in decision? Naturally, this apprehension is still preserved in retention; the ego was previously involved in it and perhaps already inclined toward it by preference ... Affective motives to orient the regard also in this direction, or to reorient it, are therefore [still] present. (Husserl 292/352)

There is every likelihood that decisions rejected once will be rejected again. Decisions once made do have a future. However, the point is that they still have to be rejected even after they have been rejected, because they never stop living as decisions still to be made. They are actually decided against—but they are virtually not decided against.

As it happens, this is an old question: Can a decision opt at the same time for two incompatible future results?

Ockham

Question III of William of Ockham's *Predestination, God's Foreknowledge, and Future Contingents* asks: "Can the will, as naturally prior to the caused act, cause the opposite act at the same instant (*in eodem instanti*) at which it causes that act?" (Ockham 71).[9] Of course, the will can choose opposite courses of action at different times. However, can it will opposite acts at the same instant? Ockham thinks not, but he acribes to Duns Scotus the view that it can.

Scotus's argument, reconstructed by Ockham, runs as follows (72).[10] If a will is contingent, as free will has to be, then in principle, it is capable of choosing both sides of a contradiction. Normally, this will be expressed by the possibility that it can will one thing at one instant, and another thing at an earlier or later instant. However, consider the following peculiar thought experiment. Suppose a will only lasts for one instant! At that one instant, it can in fact will only one thing. However, it is still contingent, and that contingency must still be expressed somehow during the time of its one-instant existence. Therefore, at that very same instant, it has to be able to will the opposite thing also. In our terms, in actuality it wills one possible branch, but virtually, it wills the other possibilities at the same time.

Ockham thinks this is all absurd (73). For one thing, the capacity to will the other thing is not a real capacity; and for another thing, it could not be actualized even if it were a real capacity; and for yet another thing, it is an immediate contradiction. Ockham probably has the more reasonable view on all counts. The idea attributed to Scotus of a will that exists for a total duration of one instant is crazy. And yet, the existential instant of decision in Sartre is something like this, where all possibilities are telescoped, not in the instant of a psychological decision, but in the self-constitution of a will that makes itself in an event. In one sense, the peculiarity of Scotus' thought experiment is to construe the will without a future. However, the interesting challenge is the idea that without a future, and only without a future, the will can set in motion both sides of a contradiction. In fact, this is a radical idea of real contradictory will. The Freudian will also wills a contradiction, but in fantasy: It desires a result both ways, and may act in one way on the surface without in the least surrendering its desire and even its expectation that things will go the opposite way. However the Scotist will does not just suppress its limits by willing both ways: It is actually capable of getting its motion started in both directions—provided, in this scenario, that it does not last long enough to find one of those actual directions cut off by virtue of logical contradiction.

What do we say about this matter, given that for us the will is futural, and not instantaneous, yet the will is capable of both sides of the contradiction at once? The role of the cutoff of future time in Scotus is played for us by the role of virtual time. Though Scotus' theory and ours seem opposite—the one futureless, the other future-saturated—we both use a non-actual form of the future to articulate a will that can go in incompossible ways at the same time for good.

Husserl's next point seems at first to mitigate this. It is a "basic error of traditional logic," he says, to think that negation is as fundamental as affirmation. "A negative decision is not on par with an affirmative decision: In theoretical statements, there is nothing of denial; on the contrary, at one time they confirm a being-thus, at another a being-not-thus" (Husserl 293/353). At first, this appears to say that decisions taken will guide consciousness from then on, whereas decisions not taken have no such status. However in fact, it implies the opposite. A decision not taken is really an affirmative decision in its own right. A negated decision option is not nothing, it is a decision otherwised. Like fictions, it is juxtaposed over the same time as the apprehensions affirmed as actualities.

This is effectively Husserl's last word on decision. The last paragraph of Husserl's chapter on decision comes to an end—almost—on this point:

> All desire for verification, the desire to convince oneself again (calling up witnesses), is motivated in science and the scientific attitude by the thought that memory can deceive, that fulfillment is perhaps never entirely complete, etc. And this is no empty possibility but one that is real, a possibility which, in becoming conscious, makes everything doubtful, to a certain extent, as to its status here and now. Thus, even intuitive certainty, transformed into a habitual possess, leads again to uncertainty, to doubt, to a question. Everything becomes questionable again. (Husserl 313)

The very desire for verification makes everything become doubtful again, even in cases where complete intuitive evidence has provided apodictic certainty.

However there is one final sentence after this passage:

> Nevertheless, we still strive for incontenstable knowledge, for convictions not subject to question. (Husserl 313)

This is not good. The conclusion that every judgment returns to doubt was based on a valid argument, drawn from the retention of countermotives over time. Why ruin it with an idealistic longing for evidence that phenomenology demonstrates we cannot have?

Husserl has a motive for wanting to say this, but the weakness of the conclusion reveals the weakness of the motive. His motive is that the fate of the unified ego is tied to the unity of time-ordering, and that time-ordering is easier to locate when experience consists of non-superimposed perceptions, nonfictional intentions, unambiguous time fixing, and irrevocable decisions. When an ego is forced to keep alive several inconsistent decision options regarding what the world actually is, the ego "is disunited with itself ... it is dragged in the direction of opposing expectational tendencies" (290/350). Husserl puts the situation strongly with lots of italics. The "subject of the world" requires that "*all certainties are organized in the unity of a single certainty*: correlatively, everything which exists for me is organized *in a single world*." Therefore, "a *practical interest* hangs on every belief, every position-taking." "The judging subject is *personally concerned* if he is compelled to abandon a judicative certainty." "*Striving for consistency of judgment and for certainty is thus a characteristic which is part of the general striving of the ego for self-preservation*" (291/351).

On a simple reading, this passage contains the old saw that we strive for pure knowledge but cannot reach it. On another reading, though, it suggests the double relation to time function that decisions lead us into, namely the doubling of continuous judgments and divergent virtual possibilities; or the doubling of continuous and discontinuous futures. As Husserl's genetic description moves from the striving for knowledge, to the unanimity of epistemic synthesis, then back to experiments with further as-if possibilities, emphasis on one of these stages over the others would lead to one of three one-sided interpretations of phenomenology generally. Emphasis on striving treats phenomenology as transcendental skepticism; emphasis on final confirmation allies phenomenology with classical epistemic and humanist norms; emphasis on experimentation with virtualities turns phenomenology toward worlds of temporal otherness.

There are many other texts where Husserl accepts a lower degree of unification in the ego. For that matter, self-preservation need not be about stabilizing unity, it might be about staking one's life in risky maneuvers. Schematically, Husserl says, there exist "successive strivings" organized in related "pathways" "at any given time." In other words, decision is temporally mappable, and the whole "system" of decision is advanced or impeded as a whole. In this sense, the ego's "self-preservation," its life, is put at stake in every decision.

In my view, phenomenology does not necessarily depend on time-unity. Phenomenology's only hard principle is that there are intentional objects. After that, phenomenology just describes forms of givenness, whichever they are, and in as many variations as it finds to describe. Whether the ego turns out to be unified or not unified, whether time turns out to be fixed and ordered or not, phenomenology should be happy either way. Phenomenology cannot restrict its capacity for describing structures of consciousness just because certain imaginable structures would conflict with some classical dogma about the unified self. It is Husserl who invented this approach to philosophy, and even if he violates it a bit here, some of his discussions in this text are perfectly open to a finding that indecision precedes and pervades decision. For example, in discussing questions, he says, obviously correctly, that it is always legitimate to answer, "I am undecided" (310).

In summary, putting together all of Husserl's descriptions of superimposed time and indecision on the time line, we can schematize a broad set of structures to help describe the perseverance of multiple decisional options in a decision's effective future. All we have to sacrifice is the theoretical unity of time and the practical unity of the ego.

The main thing we learn from Husserl is that decisional noemata are doubled in two ways. First, they are divided in such a way that one noema seizes an affirmative result of a decisional striving, while others seize alternative results. Second, they are divided in such a way that one noema aims at the future time that will actually fulfil the decision, and others aim at the alternative time lines in which alternative approaches would be fulfilled. We take from Husserl a rich vocabulary, and an audacious will, for describing overlapping and juxtaposed temporal noemata.

CHAPTER THREE

Heidegger: The original decision to decide

Heidegger discusses decision in seven sections of his *Beiträge*: from section 43, "Be-ing and Decision," through section 49, "Why Must Decisions Be Made."[1] There is a lot of talk in these pages about how essential decision is to history and the future, but as is typical of Heidegger, history and the future have little to do with time, or with what might happen next, or with any of the everyday decisions that we might think could project into a future in some determinate way. In some ways, Heidegger's care not to be caught up in factical plans reveals structures of the future in general, and these structures are what capture our interest here. But in other ways, Heidegger's concern separates decision from the concrete projects that decisions are supposed to be concerned with. There is a point at which Heidegger engages in decisionism without decisions and with a future that is not in time. At that point, we will have to move away from Heidegger to pursue our interests.

The anti-Husserlian premise of Heidegger's text is that consciousness is not consciousness *of* objects, as it is has the status of being between subject and object; of entering into something, but not yet; and of crossing, but not to a particular end (Heidegger *Contributions* 4–5). Thought has the form of "startled dismay" (11). Heidegger is equally suspicious of "total worldviews" and of "individual opinions" (28). These motifs make it hard to see how decisions in the everyday sense could be made, as consciousness is not fulfilled by deciding for or against particular objectivities. Nevertheless, Heidegger's thesis is that there is a more fundamental kind of decision that grounds consciousness.

Heidegger's working definition is that decision covers any act that is "retrievable and repeatable" (39). This is an odd definition of decision, but it may be a good one: any beginning that can be followed up on will count as a decision. For Heidegger, it follows that all decisions are unique. Again, this seems an odd thing to insist on, but the idea is that if an act is not unique but one of a chain of comparable acts, then it was not the first act in the series, and so was not the original decision that set the series in motion. To speak symbolically, beginning is "very far back." (41) If we push this to the limit, it means that only an act that responds to nothing other than itself can be the beginning of a decision procedure. It is certainly unusual to regard decision as an act that responds to nothing and is responsive to no needs or requirements. But there is something appealing in Heidegger's premise that a decision, if it really is decisive, is way back at the beginning.

Although the decision begins from way far back, Heidegger gives equal value to the idea that it "prepares for those who are to come" (43). In some cases, of course, a decision is rejected by those who come later. But even so, we could still say that decisions prepare for someone to refuse it. In any case, a decision is always for another to follow or not, for a relay of decision-makers, not especially for the decision-maker herself. Heidegger builds the future into decision not in what one is committed to, but in another who will take on the commitment in turn. The decision-maker herself, if there is one, is never more than just one of those for whom the decision prepares. Still, Heidegger's primary concern, even regarding the future, is not so much when the decision *will* hold, but when it *started* holding.

Heidegger's discussion begins at section 43 with the thesis that decision is the ground of being. "We must grasp [Be-ing] as the origin that de-cides gods and men in the first place and en-owns one to the other" (60). (I leave the "gods" out of my analysis.) To say it quickly and bluntly, the original decision distinguishes everyday life from the question of being, and so makes the decision-maker, Dasein, notice, and hence enter into, her relation to Being. The first decision occurs when we decide on Being. Therefore the first decision is the beginning of the history of Being. Therefore the first decision is Being's own emergence. Being is given to us in the form of a decision that binds us to time and space, Heidegger says. We own up to our decision, and we owe ourselves to it; we equally owe it to ourselves and we owe it to the decision to spend our time on it wherever we are.

In this account, decision is closely related to questioning, in that decision "opens up the free play of the time–space of its essential swaying" (60). Decision for Heidegger has almost nothing to do with human character, activities, or processes. For one thing, he treats time as a sudden event and not a gradual process. For another thing, on this account, it is Being that decides; it is not really we who decide. Decision has nothing to do with choice, resolve, calculations, preferences, courses of action, or even values; freedom has nothing to do with autonomous causality, faculties, or moral duties (61); decision is not connected with the kind of "system" (61) that assumes there is an order to our anticipations and a coherent web of representations of objects that we can survey in one viewpoint. It is difficult for me to avoid all of these factors in interpreting what it means to make a decision, but according to Heidegger, we should avoid them.

I want to press this issue. Is Heidegger's account of decision useless for the futural noema of a decision, as he seems to prefer, or can we also use his account to analyze the ground of the decisional future? Can decision be the ground not only for the anonymous decision of Being in its nonchronological origin, but also for our personal decisions at particular times and places as well? Might Heidegger's analysis serve as a model for human decision-making, even though it is so distant from an empirical description of human activities? It is possible that this use of Heidegger to understand a person's concrete decisions about what course of action to follow in the future is just not possible, as Heidegger's whole point is that decision never chooses action. Indeed, Heidegger must avoid any examples of decision that sound concrete, otherwise it would be a theory of decision and not a theory of the ground of decisions. Decision, for Heidegger's purposes, must be non-"anthropological," in order to be grounding. Admittedly, it is odd to say that the original decision is a ground when it does not describe any decision we might concretely *make*. It is odd, but perhaps there no decision-*making* in Heidegger, in spite of his preference for poiesis. This is always Heidegger's brilliancy but also his problem, namely that he detaches ontology from the ontic. But in spite of the gap between Heidegger's purposes and mine, I still want to try to read Heidegger's account of the original decision to see if there is some way to make it apply *both* to detached virtual decision *and* to ontic-engaged decision.

Heidegger's definition of the decision of Being is "the going apart, which divides, and in parting lets the en-ownment of precisely this open, in parting, come into play as the clearing for the still undecided self-sheltering-concealing" (61). The three points I draw from this passage are that decisions part, that decisions open, and that decisions remain undecided.

It is helpful to distinguish three levels of the ontology of decision: (a) Everyday decisions, which for Heidegger, are not really decisions in the deep sense: presumably this covers such things as deciding what to have for breakfast, whether to cheat on one's taxes, which party to vote for, whom to date, what book to read and what to write, whether to donate one's organs, and that sort of thing. These decisions, which for Heidegger should probably be called by some other term, are operations of the everyday social and ontic "system": the "beingness of beings"; (b) "Decision" in the deeper sense, but found in the actual world. Decisions in this sense are found wherever there is a "crossing" to the "other" sense of being, the "being *for* beings." It is a decision in favor of being, but in the context of a concrete question in the actual world. We might imagine examples where a person decides whether to blow the whistle on institutional corruption and risk his life's work, or when a nation decides whether to install a fascist leader. In Heidegger's terms, this half-heartfelt interest in Being does not thoroughly challenge the "secure interpretation of beings"; and (c) "De-cision": This is the original beginning, which "puts man in question." The second sense depends on the third: decision "arises out of de-cision" (62); the original question sets the stage for all subsequent crossings. The grounding de-cision is the decision about how to interpret decision.

In section 44, "The 'Decisions,'" Heidegger proposes that a study of the original de-cision counts as an account of history, but this will be history in a special sense. The existential task of original de-cision is to decide in favor of decision and against nondecision. This de-cision engages a risky situation, precisely in that it does not make any particular promise or set in motion any particular state of affairs. It does nothing but commit to treating being as something still to be decided on. It makes Being itself risky, as it admits only an open future. In this sense, decision remains in principle always so far a matter of indecision. In fact, one could imagine this (following Derrida; see Chapter 8) as a definition of democracy, which would give such a decision great practical import in spite of its detachment from everyday particulars. Paradoxically, the secure situations of everyday life, which do not call for any practical decisions, leave no room for in-decision, which is the authentic state of decision.

The question to be decided is therefore whether (a) Being reduces to a sense of security with a history of beings not requiring decision, or (b) Being leads us to pass on beings, and so leads us to "the other beginning of history" (62). In different terms, the question is whether Being is an answer or a question, whether decision is a determinate decision or the beginning of a sequence where the same decision will be posed incessantly in different modes. Heidegger has ten ways of articulating the same alternative: decision regarding beings versus decision regarding Being.

1 Is man an animal, a subject, a substance, a rational and cultural being; or is Dasein an "evolving site"?
2 Is Being general ontology; or is it unique word, attunement, and event?
3 Is truth defined by correctness, certainty, representation, calculation, and lived experience (the last of these terms is Husserl's, but Husserl would certainly not have identified it with calculation); or is truth sway, aletheia, and clearing/concealing?
4 Are beings average; or is each one a unique question?
5 Is art something to exhibit; or is it a work of truth?
6 Is history about pioneers and discoveries; or is it a surge of estrangement?
7 Is nature exploitable; or is the open earth unrepresentable in images?
8 Must the demystification of the past (which Heidegger regards as inevitable and fully appropriate) lead to onto-theological Christian culture; or can it lead to an undecidable close-yet-distant relation with the gods, namely to a "'space' for decision"?

9   Does thought lead to satisfaction; or to the risk of going under?
10 In summary, is decision an act or a risk?

The problem for Heidegger in every case is to separate the risk-taking, grounding decision from risk avoiding, already apparent, overdeterminate, decision options. He would like to make this distinction strongly, and to do so without rendering the grounding de-cision entirely irrelevant to the life decisions of authentic Daseins. Managing this distinction is effectively the topic of the original decision itself.

Can the two sides of the distinction be brought together if we interpret a particular decision (like the whistle-blower's or the anti-fascist's) as an enactment *of* the grounding decision? Can a particular decision coincide with the original decision, and not just be a pragmatic selection masking over the fact that everyday situations do not really call for a decision? Perhaps it might, as long as the particular decision has the following feature: namely that the particular decision's options are expansive enough that they are not yet all visible and may or may not ever become visible. A particular decision would thus share with the grounding decision the feature of operating without fixed options, without secure meaning, and without branches that are visible at the time of the decision. Husserl would have a hard time with this idea of a decision whose content is so obscure. For Husserl, the original decision begins to seem not a decision at all, but only the rhetoric of decision. But in Heidegger's favor, how could it be otherwise if there really is a first decision? (Maybe Husserl's best rejoinder is that there is no first decision.) A very first decision will not be able to count on previous decisions having set up a determinate situation or having articulated options to decide among. A primary decision to make decisions, and to make the world a place where decisions are to be made, is, like it or not, a precondition for particular decision-making. A decision is not possible for a being who does not already have a will to decide, any more than a decision is possible for a being without values or interests. Even if we do not like where Heidegger is going, how can we not agree with this?

The question, Heidegger says is, "Why must decisions be made at all?" (63). In fact, there are two questions: (a) Why should decision be transcendental rather than particular? and (b) Why should the authentic decision be prior to history?

1. Is the idea that we have to decide to decide before we can decide, like the assumption, ridiculed by Hegel, that we have to learn to swim before we get into the water? Heidegger does seem to say this, but it is not as simple as saying that general decisions precede concrete decisions, as it takes the occasion of a particular crisis to provide the urgency out of which one might side with decision in general (as it takes a broken hammer to make Dasein notice its anxiety toward Being). In this way at least, deciding to decide may coincide with, though it might mean something different than, those later decisions that we will make after we decide to decide. That gives the original decision some concreteness, insofar as it attaches to concrete decisions. However, if the coincidence of transcendental and particular decisions gives the former some of the latter's concreteness, it also gives the latter some of the former's lack of concreteness.

2. The decision in favor of what Heidegger calls the "other beginning," that is, the transcendental decision prior to "history," is, more correctly, the decision prior to "modernity's" conception of history. We cannot even say that the original decision is ancient, since the original decision is more basic than the modern division into ancient and modern. The grounding decision may make history possible in the "first" place, but that decision is in truth repeated in all cultures at all times, ancient and modern equally. That prehistorical decision would be unique in the sense that there is only one of them, but not unique in the sense of happening only at one moment in one culture's history. This grounding sense of the "history of being" speaks not of human history but about the beginning epoch prior to even the earliest history. Heidegger calls it the "history of being," but it has more to do with the "truth of being" (64). History, in Heidegger's conception, is not a sequence of different states, but consists of the fact that Being is "gifted or refused," present and/or concealed, in any given decision. A decision is just a "site" (66) for a new cut between Being and beings.

To see how hard it is to decide in favor of Being, Heidegger recommends that we look at the history of metaphysical errors (65). A whole litany of philosophical concepts claimed to begin a new age but ended up as dead ends, hollow promises, repetitions of the same old things, and differences without a difference. What looked like important particular decisions were no decisions at all. What looked like philosophy was nonphilosophy. Indeed, just because there was a sequence of false starts does not mean they even constitute a history.

Furthermore, we cannot simply turn pseudo-decisions into decisions merely by knowing what went wrong in the past. Decision does not mean improving on errors, or tinkering with generalizations to make them marginally more probable. Grounding decision must rigorously bypass all the topics and methods from which error has until now ensued. If the so-called history of error is instructive at all, the moral of the story is to avoid the "noise" of so-called "moral instruction" that some people claim to draw from history (65). This is a nice point, important for reading Heidegger as an ontologist and not a moralist: it is not always morally bad to base one's practical life on pseudo-decisions, if that means absorbing good moral instruction from a just society so as to live according to just social norms; conversely, some Daseins who head for the ground of Being are morally awful, as everyone who reads Heidegger knows. Being is not about morality, for Heidegger, so neither is de-cision.

We have said that de-cision does not inaugurate a gradual series of further steps, subdecisions, implications, by-products, or historical developments. It follows that the future of the de-cision can only be a promise to repeat it, not an unfolding that fulfils it. Heidegger does not say much about the future of the de-cision, only that "for a long time to come" (66), decisions for Dasein will look more like uncertain questions than answers. Length of future time is a metaphorical stand-in for the ontological condition. The extended length of time in which de-cision will remain the defining feature of our age does not make the future into a unified time or a unified project. De-cision is not like writing constitutional laws, which thereafter must be followed in their original intent. It is more like the opposite: decision as a living unwritten document. And this means that the epoch of de-cision has something in common with sequentiality after all, at least in that it will contain any number of approaches, experiments, reboots, and subprojects.

There may only be one de-cision, but that one de-cision can be dispersed in any number of instances, and as these instances each have their own impetus, they need have no direct communication with the others. For this reason, the epoch of the beginning consists of "fluctuating and disconnected starts unknown to one another" (66). It is interesting that the very unicity of de-cision renders its enactments disconnected. The "long time to come" thus implies fluctuation. In order to be long term, decision must leave space for another beginning that could happen at any time. There is only a long-term decision if its expressions pertain only to the short term. In this surprising sense, the long implies the short.

Paradoxically, since the fluctuation inaugurated by de-cision does not go anywhere, fluctuation can equally be called "the stillness of being" (66), and decision-makers can be called "caretakers of stillness" (13). The future is nonprogressive, or we might say, a still kind of timelessness. On the other hand, stillness requires a break with the noisy obsessions of the past, which is why we cannot "go backwards," even to so-called "valuable traditions" (66). To put it as a paradox that could be useful for the problem of symmetry and asymmetry in the philosophy of time: the future is timelessness that cannot go backwards, timelessness with a time arrow.

Heidegger does admittedly sometimes suggest a romantic picture of the "inaugural" past, as though he is advocating conservatism against modernity. He certainly describes "decision" as "guardianship" (66). But his romanticism of beginnings is not necessarily the same as conservatism. Indeed, in Heidegger's schema, reaching "far back" (41) is precisely a way to avoid "going back" (41). We should remember *so* far back that we cannot *go* back to it. In other words, no cultural state in the past is what the de-cision is for.

In summary, the decisional beginning far back implies uncertainty in the long time to come, which implies disconnected beginnings, which implies preparation for a future of stillness. The long-ago-ness of the decision entails the long-to-come-ness of the event. Precisely because de-cision is long term, the present fluctuates, and is not united by any consistent concrete project. In brief, the de-cision is long and the present is short.

In a way, this is common sense: the timescale introduced by ontological de-cision is so long that we cannot expect any personal or cultural decision, desire, or action, to affect it. All we can do is to give the decision a little push by our own movements, not knowing if our contribution will go anywhere or connect to others. In simple terms, the possibility of a decision was given with the beginning of consciousness, and only then. To do justice to that level of decision, we separate ourselves from the pressure of current needs. The implication is that a decision can only ever inaugurate a project, never get it going. On this point at least, Heidegger agrees with Sartre.

Section 45, which comes with the simple title, "The 'Decision,'" begins with a flourish—decision is "the strifing of the strife" (66). But the main topic of this section asks: who are the people to come who will some day make a decision? In this Nietzschean passage, even though a decision has no particular content to be concerned with, it may at least be concerned with "the one who" has it, an "outstanding" one (66), a kind of overman decision-master beyond the functionary procedures of mere decision-makers.

A "decision is made" when it is "experienced" by such a one as a "mandate" with "endurable power" (66). Heidegger's focus is generally on the decision force rather than on the decision-maker, so even here when pointing to the one who decides, his interest is less on the decision-maker than on whom we might call a decision experiencer or a decision inheritor. Heidegger's own phrase is pointedly ambiguous: in place of asking who makes a decision, Heidegger asks, "By what means is the decision made?" (66). Heidegger is not asking about the reasons why a decision is made, but about the means by which a decision is carried out, carried over, and conveyed. When Heidegger asks "by what means," he does not mean "for what purpose," or "by whose action"? "By means" is neither about ends nor means. The "by what means" question is meant to subvert the relation between agent and respondent, to give to the respondent the role of conferring meaning on the decision.

Heidegger's apparently simple thesis is that the decision gets its right from "the ones to come" (66), that is, by those who either "grant" the decision or "stay away" from it. In common sense terms, this is like saying that a decision is made by the ones who are for or against it. It is not that the decision is postponed in a "random, unending way," and pinned on some individuals who will arrive "later" (66). It is more that there are some people who have a more authentic relation to past and future than most. "The ones to come" can live and decide at any time, now or before or whenever, and still have the "to come" character.

In a relatively concrete passage, Heidegger distinguishes four categories of the "ones to come," measured by how directly they are responsible for the decision: (1) "Those few individuals" who prepare evental sites, for example by poetry, thought, deed, or sacrifice and (2) "Those many allied ones," who "intimate," "enact," "consent to," or "grasp" the "laws" of openness and strife (67). It is not clear whether the central difference between the first class and the second is that the latter are many rather than few, or that they apply the decision but do not invent it, or that they decide law rather than poems, or that their comportment toward decision is psychological rather than ontological, or that they extend decision into the future in different as yet unexplained ways, or all of the above. It sounds as though the first class are the elite and the second are the followers, though they both favor decision against the everyday.

Then there is a third type of person who has a relation to the decision: (3) "Those many who are interrelated" (67), who are part of the decision merely out of "common" origins, which gives them a kind of "durability." This third group sounds like the masses, who get to be part of a decision culture due to being in the right place at the right time. In practice, it may be true that followers can accept a decision out of habit or tradition, but it seems a bit surprising that Heidegger would class these *das Man* followers on the right side of decision-like characteristics.

Finally, there is something like a background population: (4) "The single ones, the few, the many." This odd group name for decision camp followers suggest that their quantity (whether many or few) is unimportant. Their consent may not even be explicit (as group (3)'s consent is), and even their sense of community involvement and common ground may be lacking. Indeed, these people seem lost. Yet even for them, something "gathers" or "grows" into a historical "people" (67). The average everyday citizens are still Dasein, after all, and even if they exist in the mode of forgetting and fleeing, some background presupposition in their existence remains decision-involved. Normally, Heidegger resists the Hegelian objective spirit, where people would gain ontological characteristics just by the movement of world spirit into their nation, but here, Heidegger seems to think that even the masses, all of them, may and even must, become a *Volk*. Heidegger does not offer much explanation of this possibility. But perhaps the idea is that some events can be decisive, and can offer a change with an opening for anyone and everyone to move into, even if they are not decisions carried out by the will of some particular number of people. The "who" question at first seemed to be about who has sufficient will-to-power to make decisions, but finally, it is more about different ways of affirming a decision that is all around all of us in the first place.

Then, after a pause, Heidegger adds a fifth category of decision masters: (5) "Singular" "people" in a "unique moment" (67). There is not much explanation of this last group, but his point is not just that each person makes their own decision, or that special people at special times make all the decisions there are. The "moment" introduces the question of "preparing" the conditions ripe for a decision event. "How is this decision prepared?" (67). Does "preparing" a decision give decision a futural momentum, a momentum for a moment? In what way does preparation intimate an "inkling of the decisions ahead" (67).

In all of his descriptions, Heidegger flirts with a temporal analysis of decisions, never quite arriving at that result. Preparation, for example, might or might not mean that time passes before something worthy of being called decision can be made. Heidegger's spin is that preparation is a state of "distress." It is hard to be sure how Heidegger knows that an original decision can never arise from a state of calm, but what he appreciates about distress is that it is the point where a problem may "accelerate."

As always, just when he starts to talk about preparing for the future, he pauses to distinguish once again between de-cision and mere problem-solving. So-called foreknowledge of a solution, or prediction of an outcome, is what Heidegger calls "noise." For Heidegger (not for me), even "world historical upheavals" are merely the description of cultural noise, resulting from the sum total of facts, desires, goals, and opinions. Knowing how a complex situation will come out depending on how it is shoved, even if such knowledge is possible (Heidegger does not state his opinion on whether it is), is not the same as making a de-cision. Sometimes Heidegger calls world-historical upheavals "events," but he never calls them "de-cisions". So-called "'world-historical' events" exhibit the "frenzy" of "machinations and numbers, not decisions" (68). At best, if the "style" of an event is such that a "people" either "gathers" or "subsists" in it, world-historical events might conceivably (although not "immediately") open up the nearness of a decision. But the danger in overstimulated situations is that confusion will lead the people to "miss" the very decision that the event might have been a preparation for.

The who–when element of decision does have a few implications for how the future of decisions look to the ones who make them. "Decision must create that time–space" for both history and the future (68). In positive terms, decision is "the joyful mission that grows into a will to found and build" (68). The presumption that joy is the mood of decision, that decisions need not be made with gritted teeth, does not immediately fit Heidegger's earlier appeal to "distress." Nor does the motif of a decision "growing into" something easily fit Heidegger's earlier rejection of decision as "carrying out" something. Heidegger clearly thinks that decision is undermined by "chaos." But perhaps he ought to have assigned some value to chaos, given his interesting view that the multiple beginnings of an epoch of decision do not interact or even know about each other.

If Heidegger's definition of decision is opposed to chaos, it is even more opposed to order, knowledge and rational will. He certainly wants to avoid the idea that decisions automatically "roll on" in a "technicized animal's" struggle for survival (68). Both the instinct for individual self-interest, and the ideology of cultural self-preservation, call for habit more than decision. It is true that in an extended sense, even a culture that "dodges" or "numbs" decisions, that is, even a culture of nondecision, still has some "direction of decision" (69). But this generous attribution of decision to organized culture only goes so far. Following cultural norms does not truly make a decision; appealing to culture is ultimately a way of "no longer wanting a goal" (Heidegger 68). After all, we cannot honestly take credit for our own cultures as though we had decided to make them what they are. Culture only adopts the "disguise" of decision, it is not the "utmost decision from within" (68).

In summary, the whole business of cultural history is a lost cause as far as the scope of decision is concerned. And yet, Heidegger proposes optimistically, "future philosophy" might "awaken distress" and start again. With this question of the future of decision, we brush once again against the temporality of decision. Will there be decision in the future? Heidegger at this point (section

46) names his topic as a dual term: "Decision (Fore-grasping)." One might imagine that this term raises the question: decision about when? But in Heidegger's text, it is immediately replaced by the question: "Decision about what?" (69). The time element never comes into its own in the content of the decision, for Heidegger; the future is never the noema of a decision. But what Heidegger figures is that obliquely, by bypassing the direct concern to shape the future, a more constant, stillness-oriented sense of futurity will be installed in Dasein's point of view. Future goals claim to aim at the future, but they limit the future to what we already desire. So whereas decisions based on backward-aiming traditions, and decisions based on forward-aiming choices, both lead to the "loss of history," decisions in favor of Being wards off the "loss of history." By excluding time from the content of the decision, decision is the source of time. That is the idea.

To put it differently, decision for Heidegger is not about choice. Choice means accepting or rejecting "pre-given" options (69). (In my view, the vocabulary of "pre-given" is overused among existentialists. Not all choices select only among options that existed before one made the choice. Often, a person invents an option at the moment they make the choice, or chooses a general path on the expectation that they will clarify a specific option later on. I think that even for Heidegger, what makes choice a poor substitute for decision is its presumption of a determinate future, rather than its pre-given availability.) Decision, in contrast, means being willing to give up what the decision itself "creates," and to preserve "in advance and beyond" the decision itself (69, 71), that is to preserve that creativity's history and its future. The self-defeating picture of the future that choice contains is that it picks a future state to aim at, but then, as Sartre would say, has no other future to resort to when that particular choice starts to go astray. The ideal for decision is not to pre-give (here, the word is right) the future, but to set a path into a future that has not yet become determinate. In this sense, the decision is "about" futural Being.

Heidegger says something very similar in the *Introduction to Metaphysics*[2]: "with de-cision, history as such begins" (*Introduction* 116). Here, de-cision for Being against Nothing means decision for value against indifference. Decision cuts into the indifferent totality of the world, and divides out, or selects (the *scheiden* in *ent-scheiden*) a finite portion of it upon which to confer significance. Not every being is of equal worth. Some contingent portion of being, even if so far it does not look like much of anything, is worth committing human history to. This suggests my idea that decision is about the commitment to something not entirely actual, and about pursuing it as if it had more reality than it actually has.

Preserving an open future is a way of preserving Being even in the chaos of beings. But now Heidegger says something unfortunate. "Saving" beings by decision, he says, means a "justifying preservation of the law and mission of the West" (69). By insinuation, the future is the West. It is too bad that Heidegger's politics blocks him from his own topic, the decision's temporality. Let us forget we ever read this passage.

In any case, the corollary of the idea that choice selects a too determinate future is that choice is an either–or disjunction (section 47). To be sure, any theory of decision has to say something about disjunction. Every "decision is decision between either–or" (70). But Heidegger goes on to ask the question that is his genius: where does the either–or form come from? For Heidegger, the either–or ontology of Being contains more than choosing between "this or that" being. Disjunction does not require that all the specific alternatives be pinned down in advance. A disjunct can include an indefinite number of disjuncts. Nor does decision mean selecting "*only* this or *only* that" (70). Not all decisions end in exclusive disjunctions, where the result of the choice is singular. The purpose of decision is not to reduce the either–or situation to a unitary situation (as Husserl sometimes simplifies it). Coming to a solution should not simply defuse the distressing fact that there ever was an either–or problem. Decision admits a third possibility beyond both either–or and one-and-only. Neither exclusive nor inclusive disjunction describes the urgency of the decision. At one point, Heidegger names the third possibility, the true resolution of decision: "indifference." This does not mean refusing to face the challenge, of course, but to allow the either–or to be resolved in dual form, as a kind of conjunction rather than disjunction.

Either–or is not this-or-that, but something like what Heidegger calls the back-and-forth "sway" of Being. After all, the decision between Being and non-Being, between authenticity and inauthenticity, between de-cision and choice, is an inclusive disjunction between origin and originated. De-cision introduces a kind of "opposition" into Being, Heidegger says, but as the "bursting cleavage of Be-ing itself" (71). Heidegger's picture looks something like the category of limit in Fichte, partitioning an indefinite ground position into continuously shaded possibilities. Maybe Heidegger means that decision articulates inherent limits to any proposition, thought, or action, or that decision marks the mutual limits of different possibilities. If this is indecision, it is of a unique sort.

In the final analysis, in some sense every decision succeeds, since no decision chooses. Decisiveness is an iterated aporia of the undecided question, to put it in a Derridean way. Or we might say that decision is about life (71), in that the original Being/Non-being impasse is always at stake no matter what the choice. Does this mean there is nothing left to decide, since, like it or not, decision is always "in favor of Being"? If the decision already is, why bother making it, why treat it as an act that some people do better than others, or as task not yet done, a task with a future?

It seems that Heidegger has prepared to say that everyone at every moment of life is engaged in deciding in favor of Being, no matter what the particular content of their consciousness. And yet, Heidegger indulges in this division of value between deciders versus nondeciders, just when he says it is wrong to do so. Those who merely seek survival have defined for themselves a criterion of good and bad decisions: namely, that good decisions lead to self-preservation and bad ones do not. This quest for mere survival encourages the "comfortable" perspective of the "low" "masses." It is obviously disingenuous for Heidegger to say that the lower classes live a life of comfort. But there is an interesting twist to Heidegger's elitism. If everyone cares about Being, then everyone is the elite, and every choice we make is an original de-cision in favor of decision. Those who care about Being will indeed regard every decision as well made. Those who care about beings are the ones who assert a hierarchy of good and bad decisions and their makers. Heidegger is not the elitist, he says, the people who criticize elitism are the elitists. And just as they assert the superiority of those who make the best decisions, they thereby show themselves to be makers of short-term decisions with flawed results. In other words, the one who asserts that some decision-makers are superior guarantee by that assertion that they themselves are inferior. There is almost a fair judgment in Heidegger's twist, except for the fact that he fairly clearly belittles the masses in his own voice.

At least, the question of the value of decision is a good one. "Why Must Decisions Come Down?"[3] (section 49, 71). It makes sense to ask "why" of a decision, only if decisions are free—that seems obvious—and therefore "active"—also obvious. But the other part of the freedom of decision, and of the meaning of asking "why" of it, is that a decision is "sequential" (71). As is so often the case, what looks like Heidegger's best point, the point that ends Part 1 of the *Beiträge*, and so ends his discussion of De-cision, is cut off almost before it gets started. Decision never quite gets its temporal exstasis.[4] In Heidegger's quasi-historical terms, activeness means that the decision extends from "'before the activity' and reaches beyond it," toward the future decision-maker to come. This description almost sounds like it could even pertain to everyday decisions, where there is a pre-event, an event, and a post-event. But Heidegger's idea is to put the decision into time, precisely by not limiting it to sequential chronology. This is the problem: what is the future without sequentiality?

In *Being and Time*, Heidegger says a few more things about the future that a theory of decision can use. He clearly announces the priority of the future[5]: "in enumerating the ecstases, we have always mentioned the future first" (*Being and Time* 378). "Anticipatory resoluteness" is *Being and Time*'s spin on decision, with its "future as a coming-towards." Decision experiences the future as if the world we have chosen is coming toward us. On the one hand, the person committed to a decision will, when enacting it in the world, come toward herself. "The futural Dasein can be its own ownmost 'as-it-already-was'" (*Being and Time* 373). Yet the arrival of the future is not smooth, since the world decided on will, as the future comes, step-by-step shift from being a virtual world to being the actual world. Insofar as a decision never gets entirely carried out, the arrival of the world from the future never entirely solidifies out of the virtual. But it is only to the extent that the future contains the actualized decision that we will ever have made a decision at all: "The character of 'having-been' arises, in a certain way, from the future" (*Being and Time* 373). Ultimately, every decision is unfinished, just because there is time. Anything at first articulated "in advance" has an unlimited range of "not yet" before it (375). It would be inauthentic to reduce the "not yet" to "not yet—but later" (*Being and Time* 375); the not-yet is the future at stake already now. The future "awakens the present" (*Being and Time* 378) in the sense that any resolute present articulates a future world to be in; and likewise, the future gives "rise" to the past (*Being and Time* 373) in the sense that the future's unfinished character gives the past decision something to still be working toward.

All of these features are important for a phenomenology of decision, particularly for the resoluteness of decision. But they still say little about how a decision plans and replans. Heidegger's rejection of the vulgar idea of "later" is a rejection of sequentiality in general, as if it matters little to a decision how it gets to be carried out. This, in my view, is the general problem with Heidegger. Despite the brief mention of sequence, of all the candidates that Heidegger considers as to what decisions might be about, the future is not one of them. In fact, studying Heidegger's discussion of decision forces us to recognize just how difficult it is to get a decision to be about the future. It might be that this is an idiosyncracy of philosophy, namely that it tends to treat its topics more or less atemporally. Or there might be something peculiar to decision, not shared by attitudes like prediction or hope, which makes it hard to be *about* the future. Heidegger says with his last words that the time–space of decision is about Being-historical Dasein, but he denies that decision is about "moral-anthropological time." For better or worse, I am interested in the moral-anthropological time of decisional futures.

In the next chapter, I look at decisionism, the theory that the value of a decision consists only in the fact that it is made decisively, and not in what kind of future it aims to bring about. If decisionism were right, then it would not be a problem that Heidegger says so little about how decisions structure the future, even while saying that decision opens the future. But that is the problem with decisionism: that it does not force us to have a theory of the future.

Is Heideggerian authenticity decisionist? We might be tempted to say yes, in that authenticity presumes that if a decision is made in the right spirit, with anxiety and resoluteness, it is a fortiori appropriate. But even for Heidegger, authenticity has more content than that. At the very least, as authenticity is inconsistent with egoism, there are constraints on what counts as an authentic decision. What makes it right is not just the fact that a person makes it "on their own," but that it follows the logic of authenticity, which includes being-with-others, transcending facticity, canceling "pre-given" actualities, questioning cultural assumptions, anxiety over one's own projects, and so on. In short, authenticity is not decisionistic, if decisionism means that all decisions are by definition right or binding. Heidegger would not want to have been a decisionist, and this is probably to his credit. But this means that the decisionist argument that we do not need a theory of the future would not have helped Heidegger.

CHAPTER FOUR

Kierkegaard: Decisionism in religion. Infinite futures

This chapter considers decisionism in religion; the following chapter will consider decisionism in politics. In blunt terms, decisionism is the thesis that the value of a decision is found in the fact that one makes it, not in its content. In that extreme form, few people would believe it. Even the most choice-affirming existentialist still thinks that one has to decide for freedom and not for escapism; even the most judge-centered philosopher of law thinks that a judge has to decide on some legal basis and not arbitrarily; even an advocate for the sovereignty of a political leader accepts that decisions can only be legitimately made by certain people in certain contexts after certain conditions are met. Decisionism cannot be totally opposed to considerations of content. Nevertheless, decisionism gives its theoretical attention, and its highest practical value, to the moment of decisiveness, and de-emphasizes the structure and value of the future that decisions lead toward.

In the context of theology, decisionism is the idea that deciding to believe in God is sufficient to have the belief; that conversion is first and foremost an act of decision; that one can decide to believe in God without having evidence for, or experience of God; that one need have no particular concept of God in order to decide to believe in God; that the decision need undergo no preparation, and need have no future effects. The faith is the decision.

Whether belief is decisionistic in this way seems to be more an issue in Christian theology than other religions. Whether there is a deep reason for this or not, I will focus primarily on Christian models in this chapter.

There may be some truth to the idea that belief involves will and not just acceptance. Moreover, it seems natural to think that a new belief can break with past beliefs, and that a conversion may be undertaken without much forethought about what comes next. However, in its extreme version, decisionism will extract the will to believe in God from any understanding of God, and from any particular way of life. I will not take a stand here about what the right way to make decisions about religion might be. My interest is in the way decisionism adds and subtracts elements of futurity from decisions.

The clearest current debate can be found in web-based evangelistic literature,[1] where the thesis of "decisionism," also known as "decisional regeneration," is contrasted with "conversion." Decisionism, as presented by its critics, is the view that deciding spontaneously to declare one's faith is sufficient to be saved. The controversy concerns the practice of "invitation," or the "call to the altar." The pastor invites congregants to come to the altar to confess faith; an individual rises, says something (often more or less predictable), perhaps speaks in tongues, and is declared saved. The critics of this practice hold that the act is too external, too formulaic, and too voluntarist. To be saved, they hold, requires long-term commitment rather than instantaneous decision, an inner state rather than the mouthing of phrases, and most important, it requires God's grace and not just the individual's will to decide. Conversion, these critics say, involves more than decision; being born again requires more than decisional regeneration.

From a classical tradition of religious humanism, this picture of the quick and clean decision to become religious does seem rather irresponsible, although who am I to say it is not good enough? For a more subtle version of decisionism in religion, we turn to Kierkegaard's treatments of decisions, promises, and oaths.

Kierkegaard

Kierkegaard's avowed intent is that the decision for Christianity should be made without calculations prepared in advance, and should be made without concern for future consequences. Decision must be made in the moment. It is not that any decision will do, so its content has a crucial role to play. To be precise, the content of religion can be rationally analyzed, but the result will be philosophy rather than religion. The leap of faith can happen only on the suspension of ontological reasoning and ethical judgment. A subject who takes that leap will certainly have a history of rational and ethical judgments, but grounds for the leap will not be found either in the subject's past, or in its future.

Past and future are not so much irrelevant, as targeted for suspension. Every aspect of the Kierkegaardian decision reverses some element of historical presumption and promises some further movements, even if it will never be able to fulfil them. So in some sense, there are futures, and future revisions of the past, inside every decision—it is just that these are not futures that can ever become, or could never have been, actual. In a way partially similar to Heidegger's, Kierkegaard's account of decision suggests a virtual decisional future at the same time as it makes the actual decisional future hopeless.

Kierkegaard's *Concluding Unscientific Postscript*[2] contains a lot of promises, sometimes equated with decisions. The promises are generally broken, and even when they seem to be fulfilled, it is because they were deceptive or ironic in the first place. As in all his works, his analysis is threaded into the folds of the prefaces and digressions that make up the book. The first few pages of the Introduction already make promises and decisions sound complicated, both to make and to fulfil. (a) Kierkegaard says he had (tentatively) promised a second volume of this book. (b) A reviewer had promised to review the book, but did not. (c) If he could choose between a positive and a negative review of his book, Kierkegaard says, he would choose the negative. A positive review obliges an author to do all sorts of boring things: To put down his cigar and give speeches, to pay the reviewer back *quid pro quo*, and so on. A negative review obliges the author to nothing. In general, and this is almost a summary of Kierkegaard's main philosophical thesis: "The negative is not an intervention; only the positive is" (Kierkegaard 8). (d) In any case, two years later, people "remembered" having read that never-written review (10). (e) Kierkegaard wonders: Will readers of this book, now that it has entered its second printing, forget the contents of the first printing and believe that this book is the fulfilment of Kiergegaard's promise of a sequel?

These twists raise a lot of problems. What is the difference, experientially or ethically, between the real fulfilment of a promise and a believed fulfilment of a promise? What exactly is the content of a promise? Is every book equivalent to a promise to write a better book? What is the nature of a sequel, an imitation, a postscript, or an "autopsy" of a postscript?

In particular, if a book or a theory sets out to do the impossible, what is it promising? After all, both faith and philosophy are impossible to complete, or to fulfil, in their own ways. One way to say that neither arrives at a conclusion is that philosophy promises a conclusion but cannot deduce it, whereas faith promises something other than a conclusion. If the promise of a book is to write about faith (as opposed to researching the solution to a philosophico-religious problem), then is it inevitable that the fulfilment of this promise will remain problematic? Alternatively, if the promise of a book is to write down a system, which cannot be done, then will that mean, "retroactively," that the book never started working on a system at all? That is, if a system has no possible conclusion, does that mean it cannot even count as a promise to have one? It seems that neither a work of faith nor a work of systematic theology can be fulfilled. But whereas a system that cannot fulfil its promise was never an honest promise in the first place, a "decision of faith" promises nothing more than a problematic fulfilment, and so at least extends its promise through its unfulfilled future.

The key point, to move to the thesis statement of Kierkegaard's book proper, is to eliminate "the quantifying introduction to the decision of faith" (15). The preceding remarks from Kierkegaard's "Introduction" should be taken as irony: The religious attitude does arise as a decision, Kierkegaard says, and a decision needs no introduction. It should be preceded by no calculations or quantitative valuations. The decision is its own "moment," as *Philosophical Fragments*[3] defines it: It is an event which satisfies no prior criteria but rather institutes its own criteria and sets its own conditions at the same moment and in the same deed in which it meets them. This is the volitional and temporal nature of what it means to "enter into" Christianity (16), which Kierkegaard equally calls the will to decision, the leap of faith, the promise, or the oath.

The factor that determines the temporal trajectory of a decision is whether the decision is made on the basis of subjective passion (out of "infinite interest, where "the decision is rooted in subjectivity", 34), or on the basis of the probability of achieving a definite result (a finite or partial interest). The paradigm case of the latter is a decision to achieve a particular object: A decision made at a particular time, valid for a particular time, and irrelevant once that time has passed whether the object has been achieved or not. The paradigm case of the former is "deciding the question" of Christianity. When a person makes a decision regarding Christianity, it is not about any particular object. It is an unlimited "decision about his eternal happiness" (21). "The leap is the category of decision" (99).

Pascal

It is interesting to compare Kierkegaard's decision about eternal happiness with Pascal's, for whom the decision in favor of Christianity is all about choosing a future. Kierkegaard and Pascal might agree that decision is a category of the future, as long as we are talking about the eternal future and not some finite stretch of temporal future. However, it is worth pausing to consider what sort of future Pascal's wager has as its noematic correlate.

Everyone knows Pascal's wager.[4] Those of us who do not know whether God exists have two options. We can choose to believe in God, and live a good and modest life: If it turns out there is a God, we get eternal happiness; if it turns out there is no God, we might have given up some fun in life, but otherwise do not lose much. Or we can choose not to believe in God, and if we are lucky, live a more fun life: If it turns out there is a God, we will be punished with eternal suffering; if it turns out there is no God, we will have gained more fun in life, but otherwise will not gain much. Of the four possibilities, only two have extreme results—the best result: Believing in God, when there is a God, which results in eternal happiness; and the worst result: Not believing in God, when there is a God, which results in eternal suffering. A reasonable person should decide in such a way as to avoid the worst result and choose the option with the potential for the best result. The best and worst results both arise if there is a God. Therefore we should make our decision assuming that there is a God. Moreover, assuming there is a God, we should believe in God if we want the optimal outcome.

Now, the problem is that the two options we have to decide between, that there is a God and that there is no God, do not have equal degrees of probability. If they did, then it would be obvious that we should choose the first option. (This assumes, of course, that we can believe or not believe something just by deciding to do so.) However if we go on the basis of the experiences we have so far in life, most of us have to admit that we have lots of evidence for things that have nothing to do with God, and not much evidence for the existence of God. In other words, the high likelihood option is that God does not exist, whereas the high gain option is that God does exist. The best result would therefore come from a decision in favor of a belief that has a low probability of being true.

In other words, we have to choose between a high likelihood option that would give only short-term happiness, and a low likelihood option that would give long-term happiness. The choice is between an option with strong evidence behind it that admits a low probability of eternal suffering, and a belief that could yield a fantastic outcome but requires basing one's life on the poorest of evidence. As far as reason is concerned, both choices are weak. However as we are forced to choose one way or the other, Pascal says, "reason does not suffer" (Pascal 119): It is not reason's fault that it has to choose one or the other irrational path, as those are all the options it has.

However, of course, there is another factor in the decision, namely the temporal factor. In addition to the division between high and low probability options, the options are distinguished by the character of their future time lines. The choice to believe in God offers an infinite time line for rewards. If all the choices were defined by outcomes lasting just one lifetime, the situation would be different. The outcomes for the non-belief in God option would offer (a) if there is no God: Living a happy life based on reason *for the length of one lifetime*, or (b) if there is a God: Unhappy afterlife lasting *for just one lifetime*. And for the belief in God option: (a) if there is no God: Living an unnecessarily constrained life based on faith in God for the length of one lifetime, or (b) if there is a God: A happy afterlife lasting for just one lifetime. Under those time lines, the gains and losses of the various options would be debatable. For comparison's sake, if the afterlife were conceived to last the length of two or three lifetimes, rational choice would lean more toward the belief in God option. The longer the amount of time predicated of the afterlife, the more the excellence or horribleness of the result would outweigh their shared unlikeliness, so the more we should lean toward the belief option. Still, the preferability of the belief option would remain finite as long as the time line of the afterlife were conceived to be finite. However, if we think of the afterlife as infinitely long, then the right option for the wager becomes evident.

The role of infinite time can be articulated in two ways. First, we might think that the decision is confirmed at the *end* of an infinite amount of time: The moment at which the life of belief proves its worth will occur not a finite amount of time after the moment of the wager, but an infinite amount of time after: "A game is on, at the other end of this infinite distance, and heads or tails will turn up" (Pascal 117).[5] Second, we might think that the decision is confirmed at the *beginning* of an infinite amount of time: Even if we could choose with an extremely high probability to have an extremely long happy life without God, lasting the length of thousands of normal lifetimes, and even if the probability of God's existence is extremely small, nevertheless the smallest possibility of an infinitely long happy afterlife will outweigh that: "Here there is an infinity of infinitely happy life to gain, one chance of gain against a finite number of chances of loss, and your stake is finite" (119). Where the results to be gained on the basis of a certain choice last an infinitely long time, the question of the likelihood of those results arriving become irrelevant. Any nonzero probability of the result outweights any imperfect counterevidence against the cause.

In short, for Pascal's wager to recommend faith definitively, the choice must aim at an infinitely long time line of decision results. The downside is that it will therefore take an infinitely long amount of time to find out whether one has made the right decision. Noematically, this makes the temporal content of the decision difficult to formulate. It is true that the choice "to believe in God" is formulated on the surface without mentioning time. Yet the decision that values eternal happiness does refer to a situation one chooses to bring about, and refers to the future one chooses to bring it about in. Of course, if a person decides to believe in God, but later changes his or her mind, then eternal happiness is off. The decision only works for the infinite future if through infinite time, it is never countermanded. Or in positive terms, the decision only has its effect at an infinite number of times if the same decision is made at each of those infinite number of times. The infinite decision only works if it does not branch. That is, the decision presupposes an infinite number of moments all continuing in the same direction without divergence. It will be a decision to repeat an infinite number of choices the same way. Infinite time homogenizes the difference between times, rendering the question of "when" almost irrelevant. As it always does, the mode of eternity makes it unclear whether there is any difference between times, and hence whether the concept of the future remains viable at all. To be sure, Pascal's wager may still show that the belief option is the most rational. However, because its future is unending, it removes many of the usual features of future temporality from decision-making: Sequential branching, diverging and converging anticipations, interacting commitments, layers of time scales, and so on. Again, it is not a refutation, but an interesting feature, of Pascal's wager that so much of it depends on the single issue of time scale, and that the criteria for the decision to believe in God are not transferable to any finite decision. For Pascal, finite decisions are not of particular interest anyways. For an existentialist, one might expect finite decisions to play a more significant role, but the infinite time scale of religious decision makes finite decision equally impossible for Kierkegaard.

At any rate, the appeal to the infinite future does not do away with the original feature of low probability that such an appeal started with. The eschatological picture is that we decide now, start enacting the decision right away, and receive a reward or punishment outcome much, much later, after the whole of lifetime is over, in a different time frame that leaps over lifetime. However, no amount of desirability of this outcome, and no amount of optimism, can entirely remove the doubt about whether the desired outcome set in the infinite future will be actualized. Even if it is certain that a choice to believe in God is the best way to manage the wager, the uncertainty about whether the future will confirm that choice is in no degree diminished.

Kant's postulate of hope

In fact, infinite time is a dangerous ally. In the *Critique of Practical Reason*, Kant lays out the "Postulate of the Immortality of the Soul" as one of the conditions that make it possible to aim at the highest good.[6] Kant says that to pursue the highest good, I have to believe in the "fitness of intentions to the moral law," which amounts to believing that my will is "holy," and that in principle, good will is capable of bringing about a kingdom of ends on earth. However, I know that I myself am limited to "the world of sense," so I know that in relation to this aim, I am not "at any time capable." The only way to believe in the fitness of the will is to think of it *not* "at any time," but through the length of "infinite progress" (*CPrR* 126). However that is no good either, as again, being in the world of sense, I cannot experience, and so cannot think as an object, the infinity of time. Temporal infinity by itself leaves me only with "an unattainable destination," a goal merely "hoped-for" (*CPrR* 127). Another layer of presuppositions is necessary: "Infinite progress is possible only under the presupposition of an infinitely enduring existence and personality of the same rational being; this is called the immortality of the soul" (*CPrR* 127). That is, to think that my will can aim at the highest good, given that it will take infinite time, I have to believe that I can endure for an infinite amount of time. However, of course that is not sufficient either, as again, being finite, I cannot think my immortality in infinite time using my own experience of time. So yet a further layer of presuppositions is necessary: I need to presuppose an "Infinite Being, to whom the temporal condition is nothing, [and who] sees in this series, which is for us without end, a whole conformable to the moral law" (*CPrR* 127). As I cannot think infinite time, I let God think it for me. And on the presupposition that He does so, I can then go back and assume the fitness of my intentions, which moral experience started with.

Does Kant think that these presuppositions are reasonable, or at least plausible? He certainly does not. His chapter on the Postulates of Practical Reason appears in book 2 of the text, "The Dialectic of Pure Practical Reason", not in book 1 on the "Analytic." Readers sometimes forget this, particularly in the second and third critiques, but Kant divides all three critiques into two books: The Analytic describes legitimate *a priori* conditions for experiences we actually have, whereas the Dialectic describes illusions generated by the attempt to extend our knowledge beyond what we can experience. The first page of the Dialectic of the *Critique of Practical Reason* announces its topic as the:

> unavoidable illusion [that] arises from the application of the rational idea of the totality of conditions (and thus of

the unconditioned) to appearances as if they were things-in-themselves (for this is the way in which they are considered in default of a warning critique) ... Reason is thus forced to investigate this illusion, to find out how it arises and how it can be removed. (Kant *CPrR* 111)

For Kant, fitness for the highest good, infinite progress, the idea that God sees the whole of time, and hope itself, are all illusions. Legitimate morality is not about hope, it is about duty.

The story Kant tells about assigning the outcome of moral intentions to the end of infinite time, is at least as rigorously laid out as the comparable assignations of Pascal and Kierkegaard. However the fact that Kant insists that his story is illusory lends the reliance on infinite time a spot of worry for pure practical decision-makers.

The Egyptian book of the dead

We can also see this combination of expectation and residual doubt in the *almost* unanimous ancient Egyptian ideal of afterlife. When a person in the Middle Kingdom buys a papyrus copy of the *Book of the Dead* (more accurately translated as *The Book of Coming Forth By Day*) in order to place it in his sarcophagus after he dies, his affirmation of the afterlife is made now, but the becoming-soul is awarded in a later epoch. One form of time ends for another to begin. Different Egyptian texts on the afterlife present this picture in different literary forms: As instructions for entering the afterlife, as moral teachings about how to deserve an afterlife, or as speculations about the nature of the soul in the afterlife, but never as a wager under uncertainty. Still, culturally, the matter is complicated. Even in ancient Egypt, uncertainties were expressed about whether there really is such an afterlife, and Egyptian writings survive that advocate hedonism in the present.

> I have heard the words of Inhotep and Hardadef [sages of the Old Kingdom who speculated about the blessed afterlife],
> Whose sayings are recited whole.
> What of their places?
> Their walls have crumbled,
> Their places are gone,
> As though they had never been!

None comes from there,
To tell of their state,
To tell of their needs,
To calm our hearts,
Until we go where they have gone![7]

The fact that such texts survive suggests that multiple copies had been inscribed, passed on, and placed in libraries (called "houses of life"). And if doubt regarding the afterlife is part of a culture's debate, then decisions are understood to have only potential futures, hoped-for futures, rather than definite ones. We see this sort of problematized future in the ancient Egyptian text dealing with the option of suicide: "A Dispute Between a Man and his *Ba*" (Lichtheim 195). It is not so much that the suicide is wrong because he will lose out on his afterlife—the uncertainty over the actual future cannot be conclusively overcome. What he loses by committing suicide, by ending with the present, is the promise of uncertainty itself, the future. A man who abandons his soul abandons decision.

For Pascal, and for the occasional skeptics in ancient Egypt, uncertainty about the actual future is a hard fact, even if uncertainty about the actual future does not reduce the appropriateness of aiming for it. For Kierkegaard, the role of actual time is at best a distraction away from the decision in favor of Christianity. Kierkegaard's rejection of the relevance of actual time begins as a devaluation of the past, but it implies also a devaluation of the future.

As far as Kierkegaard is concerned, when scholars interpret Christianity as a historical phenomenon, they succeeded at best in giving the stories of Jesus "historical reliability" (Kierkegaard 96), but they thereby miss the moment of decision. It is not just that evidence for what happened in the actual past is by nature probabilistic only. It is that waiting for more evidence to build up makes it seem natural and right that the "decision is postponed" (27). Dependence on the actual past not only fails to articulate Christianity's eternal future, it also presents the future as a time when historical research will carry on indefinitely. For Kierkegaard, the future, if it means anything to Christianity, is the time to make a decision about *now*. The historical future is a misleading substitute for the volitional future; the short-term future is a postponement of the eternal future; existence in one kind of time postpones existence in the next kind of time; historical events postpone the advent of true time. It is not quite as though the decision for Christianity does not exist in time at all. In fact, Kierkegaard insists that it is a "decision in time" (95). Yet as it posits the eternal, it is in that sense an "eternal decision" (96). This is one of the paradoxes that for Kierkegaard makes Christianity the exemplar of absurdist metaphysics. The most absurd and therefore the most true.

As decision for Kierkegaard means subjectivity and nothing but, the introduction of even the slightest detail of content, the merest mention of a "case in point," is destructive of decision-making. Even considering the future of one's own life "evades the pain and crisis of decision" (129). To put it bluntly, subjects can never decide anything about objects. They certainly cannot care if their ethical choices will have any effect on others, let alone on "world history" (138). Indeed, Kierkegaard thinks of world history not as something that individuals contribute to the future of, but as something that "squanders" individual decisions. No doubt there is some truth to the generalization that individual decisions are diluted, even squandered, by the sweep of history. However as Kierkegaard intends it, decision should withdraw from all consideration of history, and *a fortiori* from all consideration of world future. Ethical decision has nothing to do with how we use past time or control future time (164). Ethics is not like passing a test quickly, or like getting past a boring episode, or like doing anything fast. Its temporal value is "restraint," not speed (164).

One of Kierkegaard's theses is thus that the subject matter of decision is played out in atemporal subjectivity rather than in future changes to an objective situation. However his other thesis, equally important, is that decisions cannot endlessly be deferred into the future. It is not important to get to the future, but it is equally important not to delay the future. Pascal too requires that the decision to believe in God not be endlessly deferred, in the sense that a person had better make that decision while they are still alive. However, Kierkegaard is less likely to think of the decision to believe as a decision whose goal will only be reached at the infinite end of time. Kierkegaard criticizes postponement because the outcome of the decision will arrive right at the moment of its act.

This is just the point where Kierkegaard's irony feeds back against his theory. The delay he rejects in the theory of decision, his own decisions carry out in practice. To the principle that historical scholarship merely "postpones the decision" to take the leap of faith (27), there corresponds the prologue that postponed the writing of the text that asserts that principle. Kierkegaard's faith, his theses, and his theory of decision all become hermetically undecidable. I would not make too much of this, or to appear confident that irony outweighs the intentionality of religious decision. Still and all, decisionism shares with ironism the separation of decision from the future.

This problem presses in a long footnote of *Philosophical Fragments* where Kierkegaard makes a point of denying that a decision, once made, can ever be reversed—or even that the moment in which a certain decision might be made, once missed, can ever be recovered (*Fragments* 16). His general point, as we have seen, is that the decision for Christianity cannot be postponed; it must be made now. To put it concretely: If God offers someone a moment to accept His offer, and she does not accept it then and there, she should not assume that the offer will stand later on, or will ever be made again. If she says No to God now, and at a later date says Yes, she should not be surprised that the subsequent Yes has no meaning, and that God will reject her ever afterwards. This seems harsh, but Kierkegaard's point is that this is what it means to make a decision. A moment only has "decisive significance" under these conditions. A decision must be a genuine turning point, which moves in one direction or another. If the decision does not decide for good, he thinks, then no moment is really decisive. Indeed, if there are no once-and-for-all decisions, then there are strictly speaking no "moments," there is only an indefinite stretch of time in which the same set of options and meanings persist without change and without ever calling for decision. And if there is no moment, there is no leap into eternity either.

This is an interesting view of the "moment," a commonsense view of "decisiveness," and a painful view of religion. What I want to focus on are the two secular analogies that Kierkegaard gives to make plausible his thesis that God cannot be expected to keep his offer open beyond the moment. First analogy:

> If a child who has received the gift of a little money—enough to be able to buy either a good book, for example, or one toy, for both cost the same—buys the toy, can he use the same money to buy the book? By no means, for now the money has been spent.

But he may go to the bookseller and ask him if he will exchange the book for the toy. Suppose the bookseller answers: My dear child, your toy is worthless; it is certainly true that when you still had the money you could have bought the book just as well as the toy, but the awkward thing about a toy is that once it is purchased it has lost all value. (Kierkegaard *Fragments* 16)

The child might think this strange and unfair, Kierkegaard says, but it is normal. Kierkegaard's second analogy makes the same point. A knight is invited to join two different armies; he chooses one side, but gets captured; she then asks the other side to hire her; obviously, her time of choice is past, and the other side will tell her so.

For Kierkegaard, these two secular cases show that a decision takes place at a single moment only, and that it selects one alternative only. The alternative, Kierkegaard says, would be very strange. If a person could select one alternative on one day, then select the other alternative the next day, it would be as if she had selected both from the beginning: "If the moment did not have decisive significance, then the child, after all, must indeed have bought the book and merely been ignorant of it, mistakenly thinking that he had bought the toy" (*Fragments* 16–17). This is an amazing way to put it, but interesting. If a person could change a decision later, then it is an illusion that there has been a decision at all. By Kierkegaard's logic, if a person chose A at time-1 and chose B at time-2, then it follows that she chose B at time-1. Kierkegaard is no doubt making a little joke by putting it so polemically, but his point seems serious: A decision that can be easily reversed is hardly a decision at all.

I say he seems serious about this, but to me, Kierkegaard's two analogies are about the least convincing he could have come up with. Every city I know has both book and toy exchanges. People change their minds all the time, hedging is an important concept, buyer's remorse is both morally respectable and phenomenologically interesting, and pragmatically, there is almost always some contrivance one can think of to switch horses in midstream. The pompous bookseller (who, in the story, playing the voice of Kierkegaard, gets a kick out of humiliating the child for choosing the toy) will be surprised at what resourceful children can come up with. As for the claim that a soldier cannot switch sides once he is losing, there are exceptions from Cassius to Benedict Arnold to current-day politicians whose names I omit out of prudence. It would have been so easy for Kierkegaard to come up with examples that would have worked better. A child decides to have his tenth birthday party at the bowling alley and decides the next day to have his tenth birthday party at the park. A knight decides to fight a duel and takes the decision back after he is killed. In these cases, Kierkegaard's thesis that decisions belong to one moment only would work. Does he choose implausible examples because he is being ironical about the thesis of momentary decision-making?

Actually, my counterexamples do not work for Kierkegaard either. My examples are irreversible only because there is a time reference, or a time-sensitive reference, built into the content of these particular decisions, which other decisions do not have. What Kierkegaard wants is that "decisive significance" pertains to all genuine decisions, and that because they are decisions, they can be made only at a specific moment. The problem with the case of the child buying a toy is that as soon as we think about the decision's content, the decision's irreversibility becomes implausible. In the case of the child's birthday party, the decision's irreversibility is plausible only because a time reference is built into its exceptional temporal content. Kierkegaard wants the decision's irreversibility not to be due to its having a special type of content. And to make the point about decisions generally, he needs his reader not to think too hard about the content of his own examples. In short, Kierkegaard's association in this passage of decisive significance with an unrepeatable moment is decisionistic both in content and in method. In terms of content, a decision for Kierkegaard includes neither historical precedents nor time for future reevaluation, neither a prologue nor a postscript. In terms of method, a decision is to be analyzed without too close a regard for the details of its content, its procedure for achievement, or its temporal reference.

Kierkegaard writes ironically. Perhaps he seriously means that decisions have only a momentary window of opportunity, and says ironically that he delays his own decisions with prologues and postscripts. Or perhaps he means seriously that decisions are delayed, and says ironically that decisive significance belongs in a momentary fragment of time. Perhaps he intends to deconstruct both positions. I do not know. However, what I think does emerge from the secular cases he offers as analogies to religious decision is that as soon as we start thinking about the content of a decision, we cannot avoid speculating about its future follow-ups and reversals, and so we cannot avoid assigning it a complex futural reference.

A critique of its temporal values may not be a decisive refutation of decisionism. It is possible that ahistorical decisions are the only kinds there are, and even that religion or politics would be better off if people did not care about any case in point or the people who will be affected by them, and if they never changed their minds. The argument against decisionism is not persuasive when based on value, I think; it will only be persuasive if we can show there is an alternative ontology of time along which decisions can be thought.

We might wonder, to take one approach, whether the ironical delay of decision and the faithful timelessness of decision could be part of the same conception of the future, i.e., if each decision could be a layered act of future-making. We find possibilities like this in a range of Christian phenomenologies, from Sokolowski to Marion.

Sokolowski and Marion

Sokolowski aims to show how a single decision gets distributed into further decisions at future times and by different decision-makers. Whereas statements of fact "display the arrangement that already subsists, [decisions] display the arrangement that will be or that should be" (Sokolwoski 43).[8] The idea that decisions display an "arrangement" suits his example: "I will go to New Orleans." Once a decision has been made, "the action is determined; it is now only to be done" (93). The essence of decision is therefore the difference and the relation between the act of making the decision and the action of sticking to it. And there is a further division between decision and performance in that the decision of one person is often a decision about what another person should do, as in "the decision to send a subordinate soldier into danger." Decision does not just take responsibility for the other, but calls on further decision-making by others.

On the largest scale, Sokolowski describes decision in such a way that it will cover not just human decisions, but also decisions taken by God. The idea that God makes decisions sounds sufficiently odd that it is worth a pause. Don Lodzinski's "The Eternal Act"[9] articulates this idea: "God is a person who timelessly deliberates, decides to act, and executes His decision" (Lodzinski 325). The decision has to be timeless, because God has to contemplate the totality of all possible worlds. For Lodzinski, this requires a kind of pantheism, but for us, the main point is that a decision that contemplates the whole decision sequence would be less future than transtemporal.

Lodzinski challenges the inference that as God makes decisions about temporal events, God is therefore in time. He acknowledges that it might appear that God decides about sequences (like a chess player imagines both her own and her opponent's future moves), about a world with free will (where different people make their own decisions at different times), and about a physical universe whose causality is quantum (where individual particles have unpredictable movements). These considerations would suggest that God cannot make a decision all at once for the whole time line of the world, that God would have to be in the temporal world to make each decision at the right time, based on what else what happening at that time. However Lodzinski's thesis is that God can contemplate many possible worlds that already descriptively take into account all the events that will happen in that world over time, and decide among them without being inside the time line of any one of those worlds. God, in effect, decides which world to work on: For each world he attends to, the time line of that world will lead to the best outcome. And as God knows in advance what he will do as each contingent circumstance arises, He can make all his decision tweaks, as it were, in advance. (In terms of Kierkegaard's analogy, if God decided to create a toy, then decided later to create a book, he could schedule both decisions in the same moment.)

This is not a bad solution for a decision-maker who can see the whole world laid out at once. Of course, we humans do not see the whole world that way. So we do have to be there in future time, to tweak our decisions in real time.

In fact, classical theology (starting with the doctrine of "time as the moving image of eternity" in Plato's *Timaeus*) does not simply posit a dichotomy between timelessness and time. Any theology is going to have a middle ground, and the middle ground is precisely the place where the problem of free will and divine foreknowledge arises. This is too big an area for us here, but to take just one example, Luis de Molina's sixteenth-century text, *On Divine Foreknowledge*[10] proposes "middle knowledge" between God's knowledge of possibilities and his knowledge of actualities, a middle ground between atemporal knowledge and as yet undetermined action. In a broad sense, decisional thought contents are like middle knowledge, and human will is situated between coordinated possibilities stated and assessed in advance, and nonexisting futures that we already know will reshape the decisions we have committed ourselves to.

Although I want to distance the phenomenology of the future reference of human decision from speculation regarding divine decision, I nevertheless suggest one similarity: Namely that the human decision-maker projects forward through the decision-sequence from the beginning, spreading out into alternative future worlds.

When Sokolowski focuses on "serious decisions," his examples include "a decision to marry a certain person, the decision to adopt this child, or the choice to enter a particular religious community" (Sokolowski 76). A decision is serious if it "involves further decisions and choices to be made in response." Decision A is serious if it includes both itself and decision B. It is a decision only if it is not its own whole.

Sokolowski refers to the "decision" of the Holy Trinity to send the Son for our redemption. The relation between decision and its enactment is the relation between God's will and Jesus' "mission" (71). It might be interesting to define decision generally as a transformation of will into mission, but my interest is in the idea that decisions are distributed across times and agents. As Sokolowski describes it, the human part of Jesus was not in on the original "decision" (I imagine this is controversial in theology), only the "acceptance," "dedication," or "obedience" to it. The human part of Jesus, and thereby humans generally, have only the role of "maintaining the decision," or of "reaffirming it in new situations" (73).

This is a classical picture, both of religious truth in relation to human intake, and of the relation between decision and action: The divine decision is made only once, as an "eternal decision," but it is appropriated many times. Enactment does in a sense reawaken the decision, but cannot challenge it (not without replacing it with a different decision). Accepting the decision makes the previous decision one's own, and in that sense retroactively makes it actual, but it does not alter the decision, and in that sense is not retroactive. That is, "his [Jesus'] human choice was added to the divine decision as to something He himself had already made" (73). The most that can happen is that "a human choice can ... blend with that earlier choice and make it manifest in a new situation" (73). This is commonsense: Following up a decision with a confirming decision is essential to decision-making, but does not violate the original decision's manifest content. The decision is instantaneous but it takes time for enactment.

However this commonsense description conceals the more subtle implications. Once there is a second decision, the first decision undergoes some transformation. The enactment "blends" with the "already" of the decision, so the secondary decision is more than just an "outcome" (74) of something predetermined. The relation of decision to act is not merely a movement from before-to-after, but a discovery that moves from problem-to-solution, and indeed a realization that moves from supposed solution-to-next-problem. And once we admit this, we will eventually have to say not just of the reception of a gift, but also of the refusal of a gift, and all the forms of resistance that follow, that they are all equally modes of secondary decision-making built into the original decision's future.

I emphasize the role of decisional aftereffects. If there is any decision that would seem to be final, it would seem to be a decision carried out by God. But Sokolowski serializes the single decision insofar as it is attributed to God, to Jesus, and thereafter, to indefinitely many humans and their own aftereffects. In a similar way, Jean-Luc Marion[11] defines an "event" as something that by nature "begins a new series" (Marion 172). The novelty in the new series is not just added on to the event, it is the excess that the event itself sets in motion. The "decision to respond" (305) means being willing not just to give more, but to receive more, from the given than appears to be there, and to keep responding more as more and more is given. Or again, to cite Jean-Louis Chrétien, "The promise is the only event where we receive more than we can receive, by opening all the future to receive it."[12]

On first approximation, we might have thought that a sovereign, absolute, exceptional decision, on the formula of decisionism, removes the decision from its future consequences, that it values the moment of decision over its future, that decisionism says that more decision means less future. And it is true that the decision to value decision above all else frees the moment of decision both from factual evidence and from the test of happiness, that is, both from the past and the future. However, if the decision to value decision separates decision from future states of affairs, it does not separate decision from future decision. Whether through Kierkegaard's ironical sequels, or Pascal's infinitely lasting payoff, or Marion's supersaturated responsiveness, we find on closer examination that religious decisions, even interpreted decisionistically rather than consequentially, spill over into a plurality of decisional—not consequential but virtual—futures.

CHAPTER FIVE

Schmitt: Decisionism in politics. Sovereign moments. Habermas: Steering procedures and the term limits of a decision

Decisionism in Christian theology, to review it bluntly, is the position that rational evidence is not to the point when it comes to commitment or faith. The right kind of decision, unconstrained by the past and based on will alone, is necessary and sufficient for deciding to accept religion. In rough terms, this is parallel to the idea in the philosophy of law that what makes a law is that the legislature and/or the judge and/or the citizens decide that it is the law. Law need not be natural, rational, moral, or based on tradition; it just needs to be the product of the appropriate sort of decision. In both theology and law, the complaint against decisionism is that any crazy thing could result from such a decision. But there are two ways decisionists might defend themselves. They might argue that unconstrained decision-making does have some content—at the very least, it would lead to novelty in leadership and social order. Or they might argue that there is simply nothing that grounds faith and/or legal authority besides a decision. It is true that few theorists in any field call themselves decisionists; it is usually a pejorative term that opponents use to challenge voluntarism.

Nevertheless, the distinctive phenomenology of decisions makes something about decisionism right. Decision gets subjects involved in the future in a way that no other cognitive attitude does. If the moment of decision defines the branches that diverge from it, then in some sense decisions themselves, not the evidence they appeal to, are the ground of what happens next. As we will see, decisionism's rejection of the past leads it to reject the future as well, so it undermines its own initial claim to ground the future in the decision. But if there turns out to be a way for a strong theory of the moment of decision to connect the decision to future reference, without overconnecting it to past reasoning, that might be ideal.

Schmitt

Virtually every philosopher who values decision (from Heidegger[1] and Arendt[2] to Foucault and Badiou[3]) has been charged with decisionism, and the secondary literature is filled with commentators straining to help their favorite philosophers avoid the charge. In political theory, though, there is a proud defender of decisionism: Carl Schmitt (*Political Theology: Four Chapters on the Concept of Sovereignty*, 1934).[4]

The first chapter of Schmitt's book, "Definition of Sovereignty," begins: "Sovereign is he who decides on the exception" (5). In Schmitt's terminology, a judgment about how to apply an already accepted law does not count as a decision in the pure sense. Genuine decisions are by definition exceptions to the norm of judgment, and the content of a decision is always an exception to the preexisting law. Something is exceptional only in relation to what is usual, so an exceptional decision takes place in a society where there is still some degree of social and political order, but where a state of emergency requires an exceptional form of decision-making. In contrast, a command issued during a state of complete chaos would have to be called something other than decision. No doubt, even some emergency decrees are similar to normal political laws, and are not really exceptions. No doubt too, some exercises of power in otherwise normal situations are too raw or bizarre even to count as exceptions to a rule (12). For Schmitt, "decision" refers to a narrow range of phenomena, operating *between* normality and chaos. That scope is hard to make precise, but it is what Schmitt refers to when he says that "decision on the exception is a decision in the true sense of the word" (6).

Schmitt develops his theory of pure decision in large part to defend the legality of Hitler's will. However, from the start, his theory spins out of his control. His definition of sovereignty almost immediately excludes the Hitler type. And his attempt to allow for decisions that do not need to be justified either by evidence or by consequences, renders those very decisions decidedly undecision-like.

Schmitt's premise is that under normal norms, there are rules to follow, but no decisions. Liberalism wrongly assumes that all decisions could follow norms, and so fails to appreciate "the independent meaning of the decision" (6). Liberals, whether Kantian rationalists (14), or opponents of monarchs, or debunkers of miracles and the church (36 and 55), wish (hopelessly, he thinks) to prevent emergencies from ever taking place, and so to prevent sovereigns from ever making decisions.

For Schmitt, a "sovereign" is not defined as the highest constitutional authority. Someone who is simply the highest among legitimate authorities is really just a successful competitor for power, rather than a sovereign. (Schmitt disagrees with Hobbes' definition, 33). The sovereign is rather the one who steps in when those authorities constitutionally defined as highest cannot manage an emergency. That is, the sovereign is not defined by law, but by "what happens when ...." The sovereign, by definition, is always able to violate his pact with the people, since the sovereign is defined by his authority rather than by partnership (8–9). To put it differently, the sovereign is one who steps in from outside the usual range of competitors to resolve conflict among competitors.

Agamben[5] and Butler[6] complain that the unintended consequence of Schmitt's decisionism is that modern states have become willing to define almost *every* decision as an exception, so that its leaders can act without concern for law. In a sense, this is Schmitt's own trajectory. He begins with the thesis that decision occurs only in times of exception, conceding that liberal constitutions do the ruling most of the time. But Schmitt's project is to show that states always begin with exceptional novelties, that every judgment is exceptional as no rule can determine precise applications, and that liberalism is phony and sovereigntism honest. If this is all correct, then decisions intervene not only in exceptional cases, but almost always. This could mean either that the world is run by the strongest, or that today's world is colonized by authoritarians. On the other hand, maybe we should not complain that today there are too many pure decisions; maybe we should advocate that more such decisions be carried out by more agents—why not everyone?—defining themselves as sovereign.

 The question for us is whether Schmitt is wrong to define decision by the moment of intervention without genesis, and without futural reference. What is the content of a decision in Schmitt's form going to look like? It is one thing to theorize about the nature of the sovereign: the "who" of the decision. But does Schmitt's formula "he who decides" really go to the point of what a decision is? Why does decisionism care about the who and the when, rather than about the what? We might say that Hitler's suspension of the German constitution is extra-legal, but is that the essence of the decision? Why do we not define that decision in terms of its content, namely what it determines for the economy, for genocide, for the military, and so on? It is one thing to say that decision*ism* as a theory chooses to analyze decisions without regard for content, but it is another thing to treat decision as if it never has content.

 Without content, and without a future, the sovereign decision sounds like an exception not only to normal affairs, but an exception to world causality in general. But in that case, as an act without causal effect, it would be powerless by definition. We tend to think of totalitarianism as control, so it is interesting to think of totalitarian sovereignty as submission to a will that is unable to think about control. Or to put it in reverse, if decision is controlled will, then decisionism is just the opposite of a theory of decision.

 The removal of temporal content from decisions is nevertheless deliberate in Schmitt's account. On the one hand, because pure decisions are exceptional, they can never be repeated or continued, and so each decision is fixed at a singular moment in time. (Schmitt shares this idea with Kierkegaard, as we have seen.) But its momentary location does not really make the decision temporal. In Schmitt's scenario, the case for saying at a given moment that a decision is needed "cannot be anticipated, nor can one spell out what may take place in such a case …, or how it [the emergency] is to be eliminated" (6–7). The circumstances that might lead to an emergency are "unlimited," which is why a liberal constitution cannot pre-regulate or make preparations for who will step in, or for what the decision-maker should decide to do if such an emergency moment occurs. The decision can only be made by the sovereign itself, who defines what counts as an emergency, who determines whether there is an emergency at a certain moment in time, and who determines whether he is the one to step in, all at the same moment he steps in to decide something in particular. The decision is temporally circular, defining for itself that it is time for a beginning. It may start up a fresh temporal sequence afterwards. But a time line of social events does not determine when a decision should be made; the decision determines when the time line begins.

Schmitt may have a point when he says that continuous social history is broken by decisions, and in that sense, that there is no temporality in the question of when the decision starts, just as there is no general rule about what will get decided when a constitution breaks down, and no rule about how long it will take before the period of pure decision ends and political norms become operative again. Still, we should be able to say how decisions project changes over time—otherwise, what would it even mean to say that decisions intervene in normal sequences?

Usually, Schmitt discusses decisions at moments of breakdown, but sometimes he appeals to the origin of a social order as a moment of decision (He shares this point with Walter Benjamin, 10.[7]). The origin of a legal order cannot be justified in advance by the laws, since the origin institutes the laws that would supposedly have to justify it. A pure beginning can be nothing but a decision. Of course, to describe the stages required to institute a legal regime, something more like Balibar's[8] diachronic dialectic would be needed.

Temporally, in short, Schmitt connects the origin of a society— the prehistory of the lawful order—with the exceptional decision, or interim history, of the emergency. The time of the decision is thus either prior to events or between them (or in different ways, both). Both are outside normal time. Both will shape the future, but without any picture of a relation between past and future. The suspension of time institutes a pure present of the emergency period, as if the decision is not between past and future, but only a between of presence itself. If the interim present is a virtual politics between norm and norm, it is something like throwing the dice without knowing or predicting how it will all turn out. A sovereign decision-maker may as a matter of fact appeal to reasons for why it has to step in and do something, but the appeal will be rhetorical. Like the dice throw, a pure decision has no next phase built in, just a demand for immediate and unconditional acceptance. Schmitt claims, implausibly, that it is more important that a state should make some decision or other, "without delay and without appeal," and less important which decision it makes (56). In fact, it is inconsistent for Schmitt to insist on action without delay, as if there were an objectively correct time for action. If the beginning of an act does not depend on finding a justification, then there is no justification for acting at any particular moment, and for that matter, the sequence of actions following the decision can hardly have correct times either. The whole temporality of decision is dissolved, and with it collapses the need to act quickly.

On Schmitt's theory, who can be the sovereign? Anyone: A prince, the people, a priest, a federal authority, an individual state. There is no limit, criterion, entitlement, legal right, or point of competence that determines who can seize sovereignty (10–11). We can make a list of people who have acted as sovereigns at different periods of human history, but the status of the decision-maker varies at different times and places. It is not even that maximum power determines who will make a decision. Power does not ground decision; decision is the source of power. A sovereign does not have "the monopoly to coerce or to rule, but the monopoly to decide" (13). Indeed, decision may not imply any power at all. It is *only* decision that the sovereign has. This is why it is "decision in absolute purity" (13).

In Austin's theory of performatives[9], not just anyone can decide, and be automatically correct in saying, for example, that a certain couple is married, or that another person is guilty of a crime. For Austin, only an institution (the priesthood, the judicial system, or something of the like) can give a person the right to be the one to decide. But for Schmitt, institutions cannot govern the right to decide, as decision means precisely stepping in when institutions fail to operate. Schmitt is strictly decisionistic where Austin is not. So does it follow in Schmitt's version that if I say I decide, that makes me the sovereign? It is not obvious why an average person, despite having no power beyond the power to say what his decision is, cannot be a pure sovereign in Schmitt's sense, once he decides to be. And if many people say so, are there multiple sovereigns?

Against Schmitt's obvious intention, does decision schizo-ize the state instead of unifying it?

If decision is not about power or state authority, Schmitt's exceptionism begins to sound more like Jarry's pataphysics[10] than Hobbes' sovereigntism, more like dadaism than realpolitik. Schmitt's thesis is that "rule derives from exception" (the exception does not prove the rule, it installs the rule), and that exceptions are the mainstay of "real life." By contrast, state power is merely "mechanism ..., torpid by repetition" (15), a sorry substitute for the intensity of exceptions (Schmitt cites Kierkegaard).

To be sure, as if to compromise with common sense, there are places where Schmitt associates the sovereign with the state (12, 22, and 36), where monarchs make decisions based on their personal will for the course of a lifetime of rule. But it is not clear how Schmitt can appeal to such examples. His principle of ungrounded exception can only truly apply to someone who removes himself from legal reasoning. Nor can Schmitt assume that a sovereign decision lasts any longer than the sovereign is speaking. There is no time limit on how long the sovereign has before he has to finish his decision, or make the next one. Decision contracts to the vanishing present.

"Decision in absolute purity" does have some juristic features, because it describes the short present between past and future normality, and is meant to interrupt what would otherwise become an endless period of either chaos or stagnation. In practice, there may even be some legal precedents for the sovereign to appeal to—something like a War Measures Act. But even here, the sovereign has "monopoly over this last decision," deciding on his own terms which precedents apply (13). In other words, the sovereign treats the law as "situational" rather than "natural."

The question of opposition to the sovereign does not come up in Schmitt's text, but in practice, of course, there will always be opponents to a seizure of power, no matter how weak they may be for the moment. Even dictators have to negotiate with power factions and stakeholders. In reality, there probably never exist decisions as pure as Schmitt describes them. Even Stalin, Hitler, and Ceasar assembled backers, weakened opponents and set them against each other, faced protests and assassins and foreign pressures, manipulated public support. Schmitt generates an interesting variation on temporality and decision, but probably there exist no decisions in Schmitt's sense.

For that matter, since Schmitt's decisions have nothing to do with reasons, they will hardly be recognizable as decisions by Decision Theory. Schmitt's decisions (a) are not Bayesian, and are not based on value assessments; (b) do not aim at optimal consequences; (c) are not sequential in operation or calculation; (d) do not recognize the existence of competitors or zero sum distributions (indeed, it is odd for Schmitt, the theorist of enemies, not to consider decision-makers as competitors); (e) have no base rate conditions, since they start with situations during which nobody knows what is happening (i.e., under states of complete ignorance).

Ultimately, Schmitt says very little about what a decision-maker does, or how, or what the contents of decisions are, or how decisions become decisions, or how decisions make anything happen.

The second chapter of Schmitt's text, "The Problem of Sovereignty as the Problem of the Legal Form and of the Decision," aims to keep politics from interfering in a nation's decision-making. Schmitt is anti-traditionalist in thinking that neither the sovereign nor anyone else inherits norms from the past. Indeed, in his view, the past is never really a tradition, as the past itself was decided and not normalized. Of course, most reactionaries and fascists are anti-traditionalist in this sense. At least relative to the recent past, they advocate a break, although often for the sake of valorizing a more distant, idealized, or mythical past. But even as an anti-traditionalist, Schmitt would like to find precedents in the past of other anti-traditional decision-makers. He hunts for historical cases of pure decision, past events that had no past themselves. Insofar as Schmitt suggests that the past did have pure decisions in it, he is then after all a traditionalist of a sort. Still, in his view, modern day liberal morality, with its appeal to normalized legal tradition, is paradoxically more revisionist than his own break with the history of norms, precisely because liberalism fails to recognize how anomalous liberal norms actually are in human history. Modern attitudes toward history, he figures, are too committed to a comfortable view of history, and so cannot see those crucial moments that introduced ahistorical decisions into history.

In other words, normal history, from the perspective of normal historians, looks more sequential than pure decision does. But on Schmitt's account, it is normal history that suppresses its past.

Normal history fails to account for how the present passes in the first place. Schmitt's theory of decision eschews temporal sequence, but by doing so, he thinks, it tells a truer historical narrative.

The case of Schmitt shows that in order to interpret theories of decision, we have to distinguish between affirming the prehistory, the moment of enunciation, and the post-history of decisions. We might say that Heidegger emphasizes the decision's far past; Schmitt emphasizes the decisive moment; and Sartre emphasizes the future that decision cannot control. The past, present, and future are obviously equally essential features of time, but they can still conflict in decision-making, when a beginning cuts itself off from past and future, or in the attitude that nothing that follows a revolution can be pure enough to be faithful to it.

Schmitt's view, in short, is that he is not really more authoritarian than other philosophers, just more honest. All authoritarians think this way. Still, Schmitt has a point in noting that the idea of the "frictionless functioning" of rational calculation, carried out by professional lawyers and bureaucrats, is authoritarian in its own way (28). What Schmitt thinks his fellow authoritarians miss is the fact that every legal judgment, big or little, is a decision. Since it is not strictly derivable from the legal rule plus the facts of the case, a decision is "transformational," not computational (31). Indeed, since decision is not derived from the facts of the case, it is even "indifferent" to the case (30), "independent of the correctness of its content." The decision "becomes instantly independent of argumentative substantiation, and receives an autonomous value" (31).

Cardozo and legal decisionism

Legal decisionism, the thesis that the law of a nation is not the code passed by the legislature but the decisions made by individual judges case by case, a variation on legal realism, has as its most famous, though qualified, proponent the US Supreme Court Justice Benjamin Cardozo.

Cardozo's theory of law[11] is complicated, since he focuses not only on judges as lawmakers, but also on judges' appeal to justice in making decisions. The issue relevant to us is Cardozo's treatment of the way legal judgments set "precedents for the future" (6). "The sentence of today will make the right and wrong of tomorrow" (12). It is not that judges should, or even could, "tie the hands of successors" (95). For essential reasons, there exist "gaps in the law," variations on legal situations that laws cannot possibly anticipate, so as a Sartrian lawyer might say, future judges cannot be bound by fate, by authority, or even by good will. Ideally, though, future modification of past law will be gradual, as for example the way that property law gradually adapts to new types of economic exchange. Times change, so we have to imagine "what [the legislator a century ago] should have willed if he had known what our present conditions would be" (49). Solutions should be moderate, but guided by justice and pragmatism. "Not the origin, but the goal, is the main thing" (63). In summary, judgment for the sake of the future should be guided more by "principles of right" (*Rechtssätze*) than by "decision calculus" (*Entsheidungsnormen*) (64).

There is certainly a tension in Cardozo's combination of positions. His appeal to higher justice, his appeal to unbound judges, and his appeal to future progress seem like very different approaches. But the binding feature is that no piece of legislature can be perfectly just for all cases; therefore future judges will have to make their own decisions, and ideally, that will lead to the laws becoming progressively updated.

One obvious problem with Cardozo's seemingly liberal view is that judgments of right do not always progress in a positive way with the passage of time. One of his examples, supposedly showing the progress of modern life, is his observation that we today (1921) are more prepared to challenge the character of victims in rape cases. In previous times, he says, we naively assumed witnesses all had good characters, but as we have now learned better, the law has changed for, he says, the better.

In any case, Cardozo's position on legal realism is qualified. As he sees it, it had become a common view around 1920 that there is no law at all except the immediate decision of a judge in a particular case. On that view, general rules are not law; past decisions are not law; and even present decisions are not law for anyone but the litigants in that particular case (75). ("There are no such things as rules or principles, there are only isolated dooms," 75–6.) On that theory, "law never is, but is always about to be" (75). Cardozo finds this formulation too extreme to be realistic. It may be true in cases where there are gaps in the law, but "often there are no gaps" (76).

In some ways, he says, it is too bad that neither legislators nor judges can predict what laws or precedents to make, and that judges have to decide cases only after crimes have been committed, and cannot prevent bad acts. "We have to pay in countless ways for the absence of prophetic vision" (90). ("Hardship must at times result from postponement of the rule of action till a time when action is complete," 91.) Therefore, judges have to make decisions "retrospectively" (91). But then how do citizens know what the law requires of them?

Because there is no definitive answer to this question, the most important responsibility of a judge is not to create unnecessary hardships for the people who could not have guessed what the law would require of them. It is inevitable that even in the same courtroom, a different judge might decide the same case differently a week later (93). Even though precedents are not absolute, and ultimately are barely reasonable, for the sake of the social good we should not abandon them altogether. And for pragmatic reasons, "the labor of judges would be increased almost to the breaking point if every past decision could be reopened in every case" (93). For this pragmatic purpose, even if it violates the nature of both justice and time, law does have to be based on the past, at least in a "relaxed" way (93).

For all the difficulties of legal decision-making, Cardozo concludes that maybe we should not "worry ourselves overmuch about the enduring consequences of our errors ... In the end ... the future takes care of such things ... [and exhibits] a constant retention of whatever is pure and sound and fine" (109).

To formulate it bluntly, the unqualified focus on decision-making over precedents defers the meaning of the decision into the future, but it makes the future unknowable and makes the law unintelligible to its subjects. A hard look at the social facts supports this view of things. But it is too hard. For practical purposes, relax. It is worth the cost to allow past decisions to affect present decisions, however irrationally, so that we can tell people that the future will take care of things. Maybe there is a nicer way to say this. But the honest, decisionistic, truth that the law will not be known until the decision is made to send somebody to jail, or not, in the future, makes the question of judicial "error" almost a contradiction in terms.

Schmitt's version of the old problem of whether there can be such a thing as judicial or sovereign error concerns the "faulty act of state." Schmitt takes up Socrates' argument with Thrasymachos in *Republic* Book 1: if we say the state acts on the basis of power alone, should we call a case when the state acts against its own interests a case of state power, or not? Can a sovereign ever meaningfully think, "I made the wrong decision," or "I picked the wrong time to decide"? With no norms, those thoughts would have nothing to appeal to. Yet it is absurd to imagine that sovereigns are always by definition satisfied with their decisions. It is fair enough for Schmitt to argue that decisions of a sovereign will always look wrong, faulty, and alien from the perspective of the old norms of democratic participation, and that they will look authoritarian from the perspective of norms of rational justification. They look faulty from the perspective of norms only because the norm is a norm, not because the decision is not a decision. However, while Schmitt thinks that the appearance that a "decision emanates from nothingness" is due only to the small-mindedness of non-sovereigns, it seems that on Schmitt's own account, decisions really do come from nothingness, and that periods of sovereign rule really are akin to "anarchy" (66). Similarly, Schmitt insists that decisions are not merely "aesthetic" (35), akin to the poet's right to say anything she wants. There may be bad poetry, but there is no erroneous poetry. Again, on Schmitt's analysis, it is not obvious why law is not analogous to poetry. In the final analysis, it is not clear that Schmitt's sovereign can answer Socrates' question any better than Thrasymachos did.

On many such points, Schmitt wants to advocate something, but cannot really do so. He seems to advocate power over norms, but decision turns out to be independent of power. He seems to back personal decision over calculation, but the sovereign role is indifferent to who takes it. He wants there to be a criterion to distinguish sovereign decisions from normal advantage-seeking, but the sovereign will not be sure when if ever he is acting as a true sovereign. It is not that Schmitt is led back to siding with norms or calculations after all, but that decision gives the sovereign just as little power, personality, or self-affirmation, as norms and calculations do.

Most important, Schmitt wants to cut free from the past in order to free the future, but cutting out the past immediately also cuts out the future. The decision that disregards precedents drawn from the past is a precedent for just those future decisions that choose to disregard it as a precedent. The conclusion I draw from the self-destruction of Schmitt's decisionism is not that there has to be a past in order for the future to have content. That might be true in law, and in one sense or another, it is hard to imagine a theory of time that consists of present and future but no past. It is hard to imagine a present that does not pass, or time without before–after relations. But my concern with decisionism is not that it devalues the role of the past. Sometimes it may be good to devalue the past in making a present decision. My concern is with decisionism's idea that decisions do not intend, precisely as their meaning, a range of possible follow-ups in the future. It is the detachment from the future that leaves the decision powerless, that renders the decision unchallengeable, and that empties out decision's meaning.

Adorno and Derrida

Adorno points out a similar contradiction in Schmittian decisionism.[12] On the one hand, decisionism claims to value the idea of freedom. But by locating decision in authority, keeping decision in the hands of the few, it "delivers the act to the automatism of dominion ... We have been taught this lesson by Hitler's Reich and its union of decisionism and social Darwinism" (Adorno 229). Unlike genuine historical human decision, which calls upon a series of acts over a lifetime, the decisionism of the instant isolates freedom in an abstract moment. And if other people have no say in the sovereign's decision, that decision will rule their future lives without exception. For most people, decision in Schmitt's sense leaves their course of life as empty of decisions as natural causality does.

Adorno is generally suspicious of the whole idea of "decisions," which he sees as oversimplified dichotomies. Decisions, he thinks, choose between affirmation and denial, rather than working within the plurality of standpoints that counter each other's tendency to overgeneralize. Yes/No decisions simultaneously constrain thought and create the illusion that thought is free from the perspectives of others. "The decisions of a bureaucracy are frequently reduced to Yes or No answers to drafts submitted to it; the bureaucratic way of thinking has become the secret model for a thought allegedly still free" (229). Adorno reimagines Schmitt's sovereign agent as an nonaccountable bureaucracy, but he has a point. Once the sovereign takes the moment of decision out of the hands of democratic negotiation, then while it is true that the sovereign is not interested in the bureaucratic regulations that execute his decision, it is also true that he is even less interested in having his decision overturned by some kind of vote. The sovereign hands over a pure decision to a decisionless executor, leaving no room for mixed or impure decisions in-between. For Adorno, this in-between is just where the temporality of decision would have to lie, in-between the timeless moment of pure decision and the deterministic sequence of bureaucratic execution, in decisions that branch into the follow-up decisions of other decision-makers.

Derrida's challenge to Schmitt in *The Politics of Friendship*[13] is the most penetrating. (We consider Derrida's argument for undecidability in Chapter 8.) For Schmitt, Derrida says, decision is purely subjective, and yet an actual subject is never competent to perform it. Schmitt's claim is that it is a pure event, but as it leaves everything unaffected, it is no event at all (68). Derrida's surprising spin is that as the subject does not decide autonomously, it must be the other who decides, and not the self. Derrida's inference is easiest to defend if we take time into the picture, that is, if we define the decision by its future, and therefore include future decision-makers in the original decision. The self who makes an absolutely sovereign decision is in a sense "passive," in the same way that, in the classical problematic of free will, we say that a person who makes a decision without being determined by any prior forces or reasons, and therefore presumably acts in a random way, is not really in control of what decision arises in their minds. In short, Schmitt's decision-maker is not in control of his mind, and since he does not control the future, or even give it a thought, other people in the future decide for him.

From the fact that Schmitt's decisions are not based on deliberations, Derrida makes another surprising inference, namely that they are made unconsciously. It does not matter for this point whether the unconscious has a Freudian sense. What matters is that absolute decisions are less autonomous than Schmitt thinks. Derrida at times seems to agree with Schmitt that decision takes place in an "instant," and Derrida spends so much time on Schmitt partly because he likes the idea of disentangling events from the apparent constraints of context. But Derrida also makes the correct, and common sense point, that decision events take place "from one instant to another" (69), and it is only because of decision's futural projection that Derrida can infer that one's own decisions are made at least in part by other people.

Habermas

The most direct alternative to decisionism might be called proceduralism. Decision Theory is one form of this, and I deal with that in the next chapter. In this chapter, I contrast Schmitt with Habermas and Luhmann, representatives of critical theory and systems theory, and then I look at some practical issues around democratic decision-making, or how decisions are made about how decisions are to be made, particularly the issue of term limits for decision-makers, and decisions about how to take into account the decision-makers of future generations.

The criticism of Schmitt's "Decisionistic legal theory" is a central part of Habermas's *Legitimation Crisis*[14] (98). Habermas's argument is that law cannot be reduced to any single decision-maker, whether an exceptional sovereign individual or a legally appointed judge, since law as a social phenomenon involves a moral judgment on the part of people both inside and outside the official legal institutions.

Habermas's criticism of decisionism is directed more at legal realism than at Schmitt's sovereign, but his reasoning applies to both. Habermas' thesis is that legal decisions depend on moral norms, not just on positive, that is, factual, legal conditions and commands. In order for a decision to count as law, it is not enough that the law explicitly formulate a rule, that the decision-makers be authorized by the state's legal institutions, that the institutions follow correct procedure, that it be executed reliably, and that no participants overstep their prescribed roles. It is not even enough if citizens consent from time to time to their institutions. What is required for law, and what decisionism ignores, is the process in which the stakeholders in the legal system are able to critique and affect the rights of institutions to act as they do. Law includes the large-scale social procedures in which members of the society discuss, negotiate, evaluate, and offer replacements for, the values and procedures embodied in its decisions.

It is an oddity of both decisionism and legal realism that they take for granted that most people do not make decisions in the strict sense, that decisions are limited to a small number of agents, who are generally the ones who decided that it is they who count as decision-makers. Legal realism prevents outsiders from asserting their right to make decisions, but then it insists that there is no ground for legitimating whether a decision counts other than what the legal decision-makers say. Habermas' view is not intended as an idealist appeal to morality. He is looking for a realist way to incorporate moral considerations into legal decision-making, by incorporating the fact that citizens think about morality whether legal authorities want them to or not. Citizens, inhabitants, people in other nations, and all sorts of individuals and communities will discuss and evaluate socially relevant decisions even if they have no official role in government. Even under authoritarian regimes, people talk. And just by doing so, they may have an effect on government one way or another. A well-run society will know how to steer those discussions toward more effective governance; a poorly organized society may be driven into chaos, or into a frenzy of suppression, by unauthorized discussions. Habermas clearly thinks people have a moral right to have an effect on lawmaking, but his emphasis is realist (in spite of his criticism of analytic philosophy's legal empiricism, 102–3), just as decisionism and legal realism meant to be. The bottom-line for a theory of decision, for Habermas, is the fact that modern societies include a plurality of sources of decision-making, reflection, zones of authority, institutional norms of agreement and the legitimation of disagreement, histories of interpersonal participation, patterns and grounds for skepticism, and concern among stakeholders with the power of the state and with each other, in short a plurality of sources and grounds for evaluating norms. Virtually all societies have included a plurality of power sources (princely, priestly, economic, intellectual, familial, etc.), but the modern world has multiplied and explicitly valued this plurality. It is therefore more essential now than ever both to "steer" the plurality in an effective and mutually acceptable way, and also to open up explicitly to challenges to legitimate authority and to the values it acts by. Of course, the modern world is not always fully pluralist in practice: in some ways, the modern world is more than ever complacent with, and more mediatized to accept, decisions without grounds (102). Even so, as soon as people start asking whether a decision "ought" to have been made, as they are sure to do even when a sovereign figure is the one who decides it, the decision's legitimacy is going to get challenged. That is why there is a legitimation crisis in modernity, and why that can be a good thing.

In short, the problem with decisionistic sovereign decision without moral norms is both utopian and scientific: Decisionism is both unjust and impossible to bring off.

In one way, legal realism and decisionism have opposite reasons for affirming authoritarian decisions. Legal realism holds that decisions cannot be challenged because they are made by a properly regulated authority. Decisionism holds that decisions cannot be challenged because they are made by an unregulated authority. Both aim to put the decision-maker's decision outside the purview of anyone else. But the former is made untenable by the existence of competing views on how to regulate who is the proper authority. The latter is made untenable by fact that lots of people will inevitably think and communicate about values, so that like it or not, the sovereign's decision will be second-guessed by other peoples' decisions. Both derive decision from the momentary "force" of decision, rather than from the issues that come up as soon as people reflect on (their own or others') decisions, namely as soon as the decision is seen in terms of values. As decisions trickle down, or up, over time, decisionism becomes a weaker and weaker theory.

In *Between Facts and Norms*,[15] Habermas offers a schematic picture of liberal politics and decisional temporality. We could loosely think of the American separation of powers into three levels of government as a distinction between three decision-making powers over time (*Between* 245). The Judiciary decides on the force of the past (constitutional law and precedents); the Legislature decides about the future; the Executive branch responds to issues day to day. Marxist and other forms of government may have other ways of dividing up the timescale of decision-making, but in any successful political system, there has to be some consistency of decision procedures and reaffirmations over time. Law needs to be consistent with the past, possible in the present, and right for "the horizon of a present future" (*Between* 198). Furthermore, the temporal dissemination of legal decisions has to be understood by all involved at the time they make or obey decisions. The principle of legal hermeneutics is that norms are "pre-understood" (*Between* 199). In this respect, decisionism and legal realism prevent law from performing its main function, namely to stabilize expectations, and to solve hard cases on appeal (*Between* 201–3).

It may be true that liberal moral discourse never allows a decision simply to stand, once and for all, and that it condemns decisions to endless dispute. But for Habermas, disputability is itself a sign of stability in social decision-making, as long as it channels social plurality into a reasonable management of future decisions. Good or bad, there is no alternative. In majority rule legislative decision-making, nothing can prevent decision-makers from changing their minds and repealing their previous decisions (as Husserl saw). A decision stands only "until further notice" (*Between* 306). "Until Further Notice" is a nice category.

Habermas is not looking for confusion; he would be happy to see a society promote long-run progress and a learning curve (*Between* 227). The way it should work is that while parties to a dispute may not know they are contributing to a search for truth, a skilled judge, listening impartially, can make a grounded decision based on the evidence she hears. This would make a grounded yet still disputable, but also not decisionistic, decision (*BFN* 231), extracting a judgment from social discourse distributed across many groups and individuals, distinct from what any one party says. Decision would take place *after* the partisan evidence is presented. This is not quite how Habermas puts it, but decision discourse is the future of evidentiary discourse. And it is the future of it again and again, if the decision is appealed again and again.

There are two directions in which we can take this topic. One direction is to pursue decision's risk-taking character. The second is the practical direction, involving such problems as term limits and the rights of future generations. The idea of term limits recognizes that electoral decisions are designed in advance to be countermanded. The idea of attributing rights to future generations assumes that people not yet born have a kind of vote in our decisions.

Luhmann

I begin with the first of these directions, using Niklas Luhmann's *Risk: A Sociological Theory*.[16] For Husserl, decision involves uncertainty; for Luhmann it involves risk. He begins with a history of maritime insurance, an early instance of "planned risk control"

(Luhmann 9). Risk analysis was designed (a) to control risk; and (b) to manage in advance the potential "regret" that arises when we fail at some decision and then retrospectively feel we might have been able to control the risk and avert the loss. Our capacity to calculate risk depends on events occurring with some degree of predictability, with statistical regularity, and in segmented sequences that make it possible to intervene at reliable points in the event's unfolding.

The intention behind risk management is to make "decisions that serve to bind time." For Luhmann, the downfall of the risk management picture of the world is that in reality, "*we cannot gain sufficient knowledge of the future*; indeed *not even of the future we generate by means of our own decision*" (12–13). Rationalists talk as though we can "immunize decision against failure," to achieve "security" (13) by means of reason, but this is pretension. Especially today, in Luhmann's assessment, there is no way to live that avoids risk. So while we can and should use whatever reason we can to minimize risk, we also have to accept risk and the unknown future as part of a new kind of rational worldview. It is fine to employ "safety experts" to help us to assess and manage risk, but such experts are first-order observers and practitioners, without much reflection, who assume overconfidently that they almost always have enough facts to control risk. They have a professional interest in coming to a decision quickly and confidently. Second-order observers, like sociologists, face up to the fact that we never have all the facts.

> The observer of a decision maker may assess the risk of the decision differently from the decision maker himself; not least of all because he himself is not located in the decision taking situation, is not exposed to the same pressure to decide, does not have to react as rapidly, and, above all, does not share in the advantages of the decision to the same degree as the decision maker himself. (Luhmann 68).

I am not sure how we could assess globally how much certainty we have in the future. It may be that Luhmann is wrong, and that at least in some spheres of life, we can predict the future pretty well. Maritime safety is probably more predictable today than it was 500 years ago. But on first glance, there is some force in Luhmann's presumption that when experts claim to predict stock markets, election results, scientific advances, and the effects of intervention on complex systems generally, there is reason for suspicion. However we calculate the overall risk of incorrect calculation, we might still accept Luhmann's premise that in the present, the future will always look uncertain, and that all our decisions are decisions under uncertainty. Once we get to the future, Luhmann thinks (probably overconfidently) the effects of our decisions will be known, and we will judge our decisions after the fact in a different light than we do now. In some cases, the future will unfold in terms that our current judgment does not anticipate at all. The interesting cases are neither entirely successful nor hopelessly wrongheaded predictions, but in-between. The interesting cases for Luhmann are those where the future event is in part determined by, or "back-coupled with," our decisions now (16 and 74).

When more than one person is involved in making a decision, the situation is further complicated by a synthesis of "two temporal contingencies" (17). And this is of course the norm. Communicating our decisions with each other sometimes leads to agreement, but sometimes highlights a surplus of options. The point is that each decision introduces a simultaneity of many actualities and many potentialities.

Luhmann has a strange view of time that makes his theory of risk more peculiar than it needs to be. For Luhmann, each social system has its own way of counting time. Printing presses, clocks, and calendars are among the modern era technologies for fixing events in a single consistent memory system (38–9). But from the premise that each era has its own way of measuring time, Luhmann concludes (a) that "everything that happens, happens simultaneously" (34); and (b) that "everything that happens does so for the first and last time" (34). The idea is that because a temporal model is built into a social system, it is as if all the various temporal patterns of social events are built into the social system together, simultaneously, from the start. It is within the realm of plausibility for Luhmann to argue that systems are "operatively closed" so that many events diverging in their effects are "synchronized" by causal connections in their "immediate past" (This may not be true, but the idea that the future is inherent in the past is arguable, 35). But it is weird to insist there is no open temporal sequence. After all, the very idea of risk presupposes a "breach in symmetry between past and future" (This too may not be true, but again the idea that the future is different from the past is arguable, 77) Luhmann compromises, saying that memory creates the appearance of temporal depth to avoid what psychologically would be the "unbearable overlapping" of our life decisions (35).

In some passages, Luhmann tries to patch up a dialectical picture of time by saying that simultaneity and non-simultaneity exist simultaneously (a position one might find in Bergson, or Bloch, or Balibar, or Deleuze). When "modern society represents the future as risk," it thereby tries to deal with non-simultaneity (future risk) in the context of simultaneity (social system). But by trying to securitize against risk, society "distracts attention" (37) from the paradox. The paradox of decision is that there is just one time to make it, but it appears differently from present and future perspectives. As the same person eventually gets to have both perspectives (Luhmann 40), and so challenges herself afterwards, and even imagines in advance how she will challenge herself later, her decision takes place in a feedback loop between risk-taking and regret. As there are so many events involved in a given causal chain, "uncertainty multiplies in proportion to the rigor of the analysis." Luhmann draws the extreme conclusion that "a practical point of view renders an analysis of causality superfluous" (41). That is the truth of the matter, structured by the total simultaneity of events mixed with the nonsimultaneous perception of events. Risk management attempts to avoid this paradoxical truth.

But what else can we do, Luhmann asks? In the big temporal picture, the present already regrets that the future will not be what we want. The present represents "the invisibility of time," the "blind spot" of our knowledge of how a system works (42). Yet the present is the only vantage point we have. And once we make a decision, the present will in part determine what the future will bring, therefore whatever we decide is what we will regret. "Risk calculation forms part of a historical machine" (42–3).

Luhmann complains that his view that risk assessment is a necessary fiction, has been misinterpreted as "decisionism" (43). He admits to holding some beliefs similar to those of decisionism. For example, he believes that the modern world gives access to so much information that even if rational principles could predict and control events, we do not have enough time to calculate rationally (44). All we can do is "decide" to act, and aim for the best. He advocates making decisions even under extreme uncertainty, since making a lot of decisions is a better recipe for active life than quietism or risk avoidance. However, he regards all of this as opening a new kind of rational practice around risk, rather than a call to arbitrary, decisionistic, affirmations.

Luhmann's compromise appeals to "time-binding forms," and suggests that there may be flexible norms for decision-making amendable to change as time goes by (54–5). He suggests a few practical cases. The law of strict liability (61), for example, is a norm for holding corporations responsible for decisions that result in damage, even when the damage could not have been predicted or controlled. Here, norms hold for decisions despite uncontrollable risk. Similarly, providing for the future is reasonable even though we cannot predict the degree of scarcity that will exist in the future (62–3). Decisions build uncertainty into the distribution of future gains, and also distribute authority for future decisions.

This is Luhmann's great point: There is a logic of non-judgment in decision that temporalizes social thought.

Luhmann would like to see more risk-taking in modern society. "Openness imposes decisions—and inevitably—the taking of risks. The system cannot remain neutral in the question of which of the two values [the society's own code or the codes of others] is to be selected, for this decision is instrumental in producing the connectability of its own operations ... This is the only way it can learn, the only way it can create order ..." (78–9). The rule for decision is "the rule of never letting an opportunity go by" (79). In short, the rule is to take a chance. Risk theory is not about reducing likely harm, but about opening up risky situations and distributing risk equitably and for a long time. Put differently, in the long run everything is temporary.

Term limits

The second direction for introducing temporality into political decision includes practical questions like term limits and the rights of future generations. Many political decisions are time sensitive.

Overlapping with time sensitivity is the question of planning, though planning is not exactly the same as deciding. We sometimes think of plans as rigid and utopian, as in planned economies. But a plan can allow for as much open revisability, and indeed for as many postponements of the final decision, as the decision-maker wants to incorporate. A plan need not be limited to a blueprint, a recipe, or a linear set of instructions in a user's manual. "Pluralistic planning theory" deals precisely with situations where there are many possible alternatives and potential results, where the various people involved have different expectations and values, and where there is too much information available to make a clear assessment. Such plans contain a mix of rational, user-friendly, reformist, and radical values.[17] The ideal is to offer a proposal that will be approved by an assortment of picky customers.

There is much discussion in current political science of the temporal rhythms of democratic decision-making. Early democratic theory had surprisingly little to say about time, focusing more on power sharing than on decisions. Aristotle's interest in decision is more about putting an end to deliberation than it is about starting up a future.[18] Machiavelli urges "adapting oneself to the times," but he means adaptation to the present, "in accord with the circumstances," rather than looking ahead to changes that might occur later.[19] Hume's political theory is mainly concerned to avoid factions and coalitions.[20] Rousseau's social contract is a decision to make further decisions, but in spite of all its inventive mechanisms for maintaining a lively popular will (moving the legislature periodically from city to city, etc.), it does not think of the future as the topic of legislation.[21]

Many practicing politicians downplay the decisional aspect of the actions they settle on. John P. Robarts, onetime Premier of Ontario, apparently once said that "whenever a decision was brought to him, he could almost toss a coin, because if there was an obvious choice, it would have been made already."[22] On this picture, there is hardly ever a political decision to be made. If there had been grounds for a decision, it would have been made by somebody else already; as it has not yet been made, many decisions will be equally good. Precisely because no decision has been made in the past, the decision has virtually already made itself, and any selection will make it work.

It is election theory in particular, not democracy theory in general, that focuses on decision. Election theory is precisely about length of terms, and term limits. The whole idea of regular elections is to hedge the bets of decisions.

Juan J. Linz's "Democracy's Time Constraints"[23] lists such time-based political topics as "the timing of elections; the time requirements of efficient and accountable government; the interaction of electoral cycles at different levels and their interference with other societal cycles; the democratic ambivalence of term limits; the time scarcities of both politicians and citizens; the temporal logics of direct democracy; the value of governmental stability, and the complexities of generational renewal" (Linz 19). Most of these topics are self-explanatory; just noticing the topic is enough to predict more or less what should be said about it. The point is that either too strict, or too lax, time constraints on decision-making can hinder democracy. Simple matters like whether meetings run too long or too short can short-circuit the entire democratic ideal. As politicians control the time line in decision-making, they often control the content of the decision itself.

Once democracy is described in this way, the complexity of most political questions implies that there is rarely enough time to decide wisely. The time to decide will almost universally be cut off too early; the best hope is to control the loss. Even so, most procedural questions in democratic theory involve the time leading up to a decision, rather than the time intended by the decisional noema. Even the issue of term limits can be taken in two ways: As the idea that politicians should only be given a limited time in which to make decisions; or as the idea that their decisions should expire after a certain number of years. To take another example, budgetary decisions in one way prioritize values for the future; and in another way set a point in the future at which the budgeted funds run out.

Philippe C. Schmitter and Javier Santiso's "Three Temporal Dimensions to the Consolidation of Democracy"[24] discusses timing and tempo in the transition to, and the consolidation of, democracies: sequences, rhythms, fears of the future, delayed responses, anticipated reactions, lag times, lead times, and so on (Schmitter and Santiso 69). In part, decisions select the "the right time." New democracies, for example, try to catch up to, and leap ahead of, other nations (74–5). In theory, reforms may have an "optimal sequence," but given the large number of simultaneous demands on political action (often it will not be possible to do just one thing at a time, or even one thing after another), decision-makers, especially collective ones, are "likely to lose control over the agenda of sequential responses" (77–8). "Tempo" measures the speed (or the delay) by which improvised reforms are consolidated (82). Prevailing wisdom says that the transition to democracy calls for abrupt economic reform with delayed political reform. But that ignores the layers of temporal rhythms in both (80).

Issues around the timescale of elections are equally layered. Dennis F. Thompson's "Election Time: Normative Implications of Temporal Properties of the Electoral Process in the United States"[25] lists three layers of electoral temporality: Elections take place at intervals (sometimes fixed, sometimes declared); citizens vote on the same day (although early voting is on the increase); and elections are irrevocable until the next one (with exceptions for recall or impeachment). We might structure democracy under the categories of periodicity, simultaneity, and finality. Each temporal structure opens a field of questions: periodicity raises the field of redistricting between elections; simultaneity needs to be mitigated by some forum for public voice in the midterm; and the relative finality of election day results makes pre-election processes like campaign fact checking and financial disclosure all the more essential.

Intentionally or not, decisions set structures for their own revisions, which begin even while the decision itself is still being made. To use this fact, decision-making could be designed to be more sensitive to the public's changes of opinion in the course of time. There are advantages and disadvantages for politics to be subject to daily changes of popular decision. Changing decisions in midstream is often treated as a symptom that democratic processes tend toward indecision; but in some ways, it is good when democracies are undecided (see Chapter 8).

Future generations

Very short-term and very long-term decisions have different futural structures. In normal politics, for example, an election with a term limit has a scheduled future with another decision at the end of it. But there are other decisions where the term limit is not just a limit on the thing decided for, but also a limit on the decision itself. For example, the decision to lease a car for a three-year term has a different structure from a decision to buy one car today and another car three years from now. The lease contract declares a limit on the decision period, a decision to make a certain kind of decision for a limited amount of time and then stop. A decision to manage one's own funeral arrangements, or to retire, or to break out of prison, or to lose one's virginity, or to give away all one's Lego, may be the last decision of that type that one can make. These decisions still have branching futures, but that is no longer their principle form. In these cases, the issue switches from the temporality of the future, to the temporality of the short term.

Very long-term decisions bring us to the issue of virtual time. If the short-term is not too long, that is, if long-term really means extended long-term, then there is no need of a special category of virtual time. But when long-term planning extends into periods where predictability loses its metric, where the people who will have to carry on the decision have not yet been born, or may not speak a current day language, or may even have evolved different bodily organs and cognitive functions, the timescale is not just in the order of later-on, but in the order of speculative imagination. There are issues with great practical import that have this sort of timescale, from the storage of nuclear waste, to investment in space travel, to saving the planet from global warming, to hoping that some of the unplanned consequences of gene splicing will be interesting. Such decisions may plan on a first series of steps, and may aim at an ultimate goal under a necessarily vague description, but they leave the bulk of their midrange future empty, or as I prefer to say, virtual. This is not something that could be avoided with better data. In the context of the long-term, virtuality is built into the concept of the future.

It is strange to say that we must leave the actual world's time line in order to have a chance to have a long-term effect on it, but if we are being practical, we must. In some situations, the entirety of a decision's noema may be virtual from beginning to end, with no expectation of actual results. In that respect, science fiction, or futurist manifestos, could be considered as decisions despite being incompossible with the actual world.

Virtual ethics (virtually the opposite of virtue ethics) are ethics that take responsibility for nonexistent beings at nonexistent times: Responsibility for far-future generations, or for extraterrestrial life forms whose subjectivity we know nothing about. Peter Szendy's *Kant in the Land of Extraterrestrials*[26] highlights the surprisingly frequent appeals to extraterrestrial life forms in Kant's ethical philosophy. If we want ethics to apply to all rational subjects, not just to the few of them most similar to those in our culture or with our biology, that is, if we want to be sure that morality is based on respect, we ought to act as though the beings for whose ends we act do not even exist.

Virtual others are not entirely nonexistent, of course, just distant enough not to count either as actual or potential selves. With this in mind, the "Future Library" in Oslo[27] commissions books by current authors, and locks those books away for a hundred years, to be read by future readers only. People today can purchase (for about 600 British pounds), as a future commodity, the right to read the book, and can bequeath that right to their descendents, but no one will be able to read the book until 2114 (when of course there may not be forests to make paper to print them on, or people to read it).

In environmental ethics, we have become used to appeals to responsibility for future generations. Nick Foton[28] summarizes the questions pertaining to contingent future persons under three headings: (1) Is our responsibility (a) to procreate more future people, or (b) to make sure that however many future people there are, they are left with adequate resources? (2) Do possible people have actual interests or only possible interests? (3) Should possible peoples' interests be counted, or "discounted," differently than actual peoples' interests.

We have become equally used to objections that ridicule the appeal to future generations.[29] Perhaps the most direct objection is the "Temporal Location Argument." This objection asserts that a person can only hold an obligation to another person who coexists in the same temporal location. It is hard to know what evidence could decide whether this is true or false. One way around this impasse is to cast the obligation to future generations in terms of "possible" persons rather than "future" persons. But this seems disingenuous: futurity is surely the key point in environmental ethics, if not in all ethics. Although there is definitely something odd about treating future generations as though their not yet existing views already counted, the appeal to future subjectivity is inevitable in decision theory.

There is something plausible about assuming that there will likely be subjects in the future, and that when they exist, they will have value. Yet at the same time, this has funny consequences, at least for utilitarian versions of the appeal to future generations. Normally, utilitarianism gives equal weight to all people's happiness. We might conclude that present and future people have exactly the same right to happiness. But if we have to take all future generations into account when deciding whether we who exist now can act for our own happiness, that is, if we have to consider all the future people who might be harmed by our actions, we will have to weigh huge numbers of future people across who knows how many future generations over a potentially endless amount of future time. This huge number of future people will be weighed against just a single generation of people living in the present. The future people will always win out over present people, by sheer numbers.

There is also the amusing "disappearing beneficiaries argument." If our generation acts badly, future people will be harmed. If we act well, future people will be benefited. But the people harmed in the first case are not likely to be the same people benefitted in the second case. It is more likely that different acts we could carry out today would each cause a different set of people to be born. So it is not that there are certain future people who will be helped or harmed depending on what we do. Our choices make different beneficiaries and victims appear and disappear. If our acting badly causes certain people to exist and suffer a bit, people who would otherwise not exist at all, utilitarianism might require in some cases that we act badly in order to allow those people at least to exist.

How should we distribute benefits across a community that includes a combination of some actual people and some virtual people? As long as we divide ourselves and future generations along the usual parameters, where we are actual and future people are virtual, where we live in actual time and they live in virtual time, then there is an impasse in communication between us and them, and their nonexistence will not count for much, relative to our existence. But if we think of ourselves already as virtual decision-makers in a multiple virtual time line, and think of each of our decisions as already branching into multiple meanings, then the problem is not one of us against a different status of them, but of how each of our decisions is already a decision in and for all of us in virtual time lines. It would be interesting to investigate what responsibilities future generations might have to us.

Lukas H. Meyer[30] grounds our moral duty toward the future on respect for the past and the present, not for the future. Meyer has no problem respecting the future goals of existing people, and even the future goals of past people. But he objects to restraining ourselves for the sake of the rights of potentially future people who may never exist at all. On the virtual ethics I envisage, the mere virtuality of nonexisting people is not a decisive limitation to the obligations we have toward them. To me, the biggest problem is that it is not easy to fix the temporal location of any of us, present or future. What happens to the concept of obligation toward future generations if that is the case?

Rawls' argument against time preference

When the future referent of a decision is connected with future generations, morality has to deal with the connection between delayed and distributed gratification. Rawls discusses this sort of thing under the heading of "time preference."[31]

Suppose we have to decide how to distribute a benefit—for example, to take most of it now, or to save most of it for twenty years from now, or to take equal amounts yearly. Suppose we also know that we will certainly still be alive and able to enjoy the benefit twenty years from now, so that we do not have to take into account the presumed greater certainty that events near the present will occur. In this case, it seems rational to distribute benefits equally across time. Now, suppose we have to decide how to distribute some commodity into future generations. First, apply Rawls' "original position" to temporal position: suppose that a decision-maker does not know whether she exists now or will instead be a person existing in a future generation. In this case again, it seems rational to distribute enjoyable goods equally between present-day people and future people, since the decision-maker does not know which she is. It may be that a person today values her own enjoyment higher than they value that of future people, but the future people will by the same token value their present enjoyments higher than they value that of past people. Indeed, a person living today may value her present enjoyments highly and discount her anticipated enjoyments in the future; but the same person in the future will value her enjoyment at that time highly and may discount the enjoyments she has had in the past. On any of these scenarios, Rawls argues, "Although any decision has to be made now, there is no ground for their using today's discount of the future rather than the future's discount of today" (Rawls 294). Recognizing this fact ought to lead members of society to "arrive at a consistent agreement from all points of view, for to acknowledge a principle of time preference is to authorize persons differently situated temporally to assess one another's claims by different weights based solely on this contingency" (294–5). And basing justice on the contingency of who one is, is patently absurd. In short, "in first principles of justice we are not allowed to treat generations differently solely on the grounds that they are earlier or later in time" (295). We have no grounds to prefer either the benefit in the near future to that in the distant future, or vice versa. There is no "time preference" for such decisions.

It is therefore reasonable for a society to decide both in favor of some "collective saving for the future" (295), and also to use up some resources now. "Saving for the future" is itself an interesting category for a theory of decision. The issues go beyond delayed gratification and future generations, to the way decisions project futures. By Rawlsian values, events distributed at different points in the future have equal force at the time of the decision. Valuing the present over the future assumes prejudicial temporal perspectives, or to twist a Husserlian term, temporal *Abschattungen*, temporal profiles or shadowings-off. Decision values are revealed once we move our original positions around temporally, assuming different temporal perspectives on the outcome event. To be sure, future events present themselves under more uncertainty due to their temporal backsides. The idea that we understand and value events better when we get temporally close to them has some obvious truth to it. Still, there is something right about Rawls' temporal nonpreference, namely that the future referent, however its unconcealedness is distributed, is the direct content of any decision. It is not that we decide now *for* later, but that we make a lot of decisions at many positions along the full time line. Rawls does not discuss branching possibilities in future content, but there is no reason why he should not. Can we make decisions on a branching model of the future? Can we take up neutralized preferences for branching sequences splitting off at various nodes along the time line? To build a working alternative to decisionism, we need all the tools of future reference: precedents and steerage mechanisms, risk distribution and term limits, communication with future generations, and unprejudiced time lines. But to make all of these work in detail, we need a logic of sequentiality in decision-making. For this, we turn to Decision Theory.

CHAPTER SIX

Decision theory: Seriality effects on decision

My plan is to use elements in Decision Theory as resources for a theory of the virtual decisional future. I will then look at three very different applications of Decision Theory: Systems Theory, economic prediction, and the subtle question of whether computer programs make decisions.

On the usual definitions, Decision Theory is about how individual decision-makers maximize expected utility, and Game Theory is about how they do so when other decision-makers are competing with them. The latter introduces some additional futural structures, but they share the idea that a decision made in the present, based on a combination of some (but almost never all) relevant data and a reasonable (but generally nonexpert) understanding of statistical probabilities, can lead with greater or lesser probability to a near or distant future that contains something at least similar to a goal valued at some degree of desirability. In terms of the stages, delays, impasses, reiterations, branches, digressions, feedback loops, revisions, halting-points, endgames, reboots, entropies, flops and flipflops, and all sorts of future pathways and counter-paths, Decision Theory and Game Theory play on a common field.

Decision Theory is not primarily concerned with the conception of the future, but with the rationality of a decision at the time it is made. Of course, it is concerned with how we value the outcomes that the decision will lead to, as the only way to evaluate a decision is by its probability of achieving that outcome. But to evaluate the rationality of different possible decisions, it holds the desired outcome steady as a given.

A central distinction in Decision Theory[1] is that although some decisions are one-off decisions designed to lead to single and discrete outcomes, others are sequential, or iterated. In those cases, distinctions among different forms of the future are built into the decision itself. Our discussion will center on these variations.

In Decision Theory, rationality is based on probable outcomes, on choosing actions that most of the time are most likely to lead to desired results. It is thus based on the law of large numbers, which means that practical rationality, even when the basis for a one-off decision, depends on a view of the world modeled on "the long run" (Peterson 71–3). The long run is in principle infinite. We bet on a single throw of the dice according to expected averages, that is, according to the distribution of results that we would expect if the dice were thrown an infinite number of times. The average is "fixed over time." We can always be wrong in a particular case as anomalies only revert to statistical normality over a theoretically infinite number of iterations. How many iterations are needed to justify the presumption that over many iterations the average will remain the same? That is precisely the question that the law of large numbers is intended to avoid. We want to make the same decision about what bet to make on each unique occasion. But we know that on a single occasion, the future convergence to the norm may not apply: A coin can turn up heads any number of times in a row. Neither the statistical norms nor recent events determine what will happen in the next roll of the dice. Both rules and facts equally lead to the "Gambler's Ruin" (72). In short, probability analysis is ultimately neither about the present decision nor about the past, but about an indefinite future.

The standard approach to Decision Theory is concerned with the rules (in some versions axiomatized) for calculating objective probabilities. The alternative approach is psychologist Daniel Kahnemann's subjective "prospect" theory, which interprets decisions through the decision-makers' often exaggerated expectations about what will happen. Prospects sound futural, although with the proviso that subjective expectation is often at odds with what will be the real future. Depending on the approach, Decision Theory may be more about ideal but rare decisions, or about flawed but typical decisions; more about the real objective future or about the virtual expected future.

Decision Theory presumes (rightly) that few if any decisions are made with full knowledge of past precedents, or of the current state of the world, or with complete foresight into, or control over, the future. When decision-makers know the limited probabilities of each outcome for each decision, the decision is made under "risk"; when they do not know these things, the decision is made under "uncertainty."

Decision Theory is not designed to tell people which values to assign to which outcomes. Most Decision Theory has the axiom (probably exaggerated) that peoples' preferences are transitive, ordered, and complete, that is, that people have consistent comparative preferences among any two or more possible outcomes. But the history of how a given person builds their preferences is not considered to be part of Decision Theory. The interesting field of the ongoing feedback between decisions and preferences is rather a topic for psychology or phenomenology.

Decision Theory is thus not essentially a prescriptive method for achieving optimal results. Even that version of Decision Theory that is most interested in rationality is just a description of what players will do, depending on whether their mode of analysis is rational or not, on whether they are self-interested or not, whether they have lots of evidence or not, and so on. Strictly speaking, it is not an empirical description, as experimental evidence shows that even well trained, self-interested decision-makers (e.g., economics graduate students, the experimental subjects of many empirical studies run by economics professors) will mix strategies incoherently, and when they do follow their training, will often get to more confused outcomes than untrained subjects (Heap 41).[2] In short, Decision Theory is a variable set of descriptions of probable outcomes for sets of decision-makers who use this or that method for decision-making.

For better or worse, Decision Theory starts with whatever value structures people happen to have already chosen. Some people, for example, do not want to accept even a relatively low risk of a bad result to achieve a good result; they are averse to risk. For those people, Decision Theory provides "Minimax" mathematics (to minimize the maximally bad option). Other people prefer a decision that maximizes their minimum gain (rather than either taking a risk to maximize their maximum gain, or avoiding risk in order to minimize their maximum loss): They are offered a "Maximin" calculus. Decision-makers can ponder for themselves whether it is more rational to avoid the worst case, or to have a chance at the best case, or some combination. Some decision theorists, it is true, want to determine which attitude leads to the best results overall, and some assume that all decision-makers do or ought to have more or less the same balance of desires. Some theorists, for example, assume that there is a universal "regret matrix" (50), according to which missing out on a great outcome will likely cause regret, even if it was not reasonable to expect such an outcome. If the regret matrix is universal, it ought to make us realize that a great outcome is worth far more than an adequate outcome. And this would make it universally reasonable to try for the great outcome even if it is significantly less likely. This sort of analysis tends to turn Decision Theory in a prescriptive direction. But regret too depends on what people value, and as far as Decision Theory proper is concerned, there is no rule for what one ought to prefer. Decision Theory should leave it up to each decision-maker to select their own version. We usually think about Decision Theory as a theory for quants, opposed to phenomenology, but at a certain level it only quantifies choices after people imagine living biographies for themselves. In this sense, Decision Theory is all about the future.

Decision Theory generates its most interesting structures when dealing with decisions that take place in multiple stages over time. I begin with a number of basic forms of sequentiality.

When there are two apparently equally desirable and probable options, we could either choose one of them and always do that, or else we could "randomize" them and do each one some of the time (Peterson 56). For a one-off decision, randomizing has no meaning. But if we have many chances at a similar decision, we can randomize. By the second try, we might find that one option is better after all. The clearest way to decide is to try one, stick with it until we learn its frequency of success, then switch or not on that basis.

Decisions are "sequential" when they are broken up into several different steps. The most commonly used example is that in a restaurant, you can simultaneously decide on all three courses as you begin the meal, or else decide sequentially, by waiting until you finish the first course before deciding on the next. Multistep decisions can sometimes be solved in advance with a "plan." The simple definition of a simple plan: "a plan specifies exactly your moves at all choice nodes that you can reach with or without the help of the 'moves' made by nature at the chance nodes" (177). A plan to drive from Pittsburgh to Toronto in January includes when the driver will exit one highway and turn onto another, and also leaves extra time for predictable but uncontrollable events like a winter blizzard south of Lake Erie. States of the situation will change at various points in a procedure; at or between these changes, we can make choices. A plan charts these possible sequences, and evaluates the results that flow from different divergences. In an important sense, I think that all decisions are sequential in some way, even those that have some one-off properties, like the decision to marry, or to go skydiving for the first time (it is only first once), or to drink just one glass of wine. Sequential procedures always imply intermediate choice nodes.

Some decision theorists assume that a decision-maker's plans, motives, understanding of options, and acts of will all exist as mental states simultaneously at a particular moment in time. Michael McDermott[3] proposes a more complex version of Decision Theory where options are not recognized at temporally fixed moments, but evolve as "complex courses of action extending into the future" (McDermott 225). Sequential decisions look forward from the time they are entered on the flow chart, but do not necessarily predecide all the decisions to come. In action theory, the problem is usually articulated as to how a choice being made now has causal effects on action later. However if we think of a plan not as a picture that a decision-maker has in mind at the moment he or she makes the decision, but as a framework extending throughout the time that the decision will be enacted, then we might think of the plan as a cause of future acts and choices.

According to the usual (phenomenologically doubtful) assumptions of Decision Theory, a decision must aim at some value as its final goal. But in-between taking a first step and achieving the final goal, Decision Theory allows for all sorts of gaps and uncertainties. After all, much is unknown at the time a decision is made, and any rational person must recognize this and plan accordingly.

For that matter, analyzing decision from the standpoint of the starting point is only one way to proceed. Aristotle's account of deliberation, for example, pictures the decision as starting with the end point and thinking a procedure backwards, until the backward analysis comes to something the decision-maker can do right now to start up the process that will lead to the end.[4] In simple terms: We imagine the best final result and work backwards to what we should do now.

What Decision Theory thus calls "Backward Induction" (Peterson 177–9) sounds like it reasons about the past, but it is about the past of a future, so the whole line of reasoning is actually about the future starting now. If a person cannot imagine themselves in the future, they cannot reason. As Philip Gerrans[5] shows, patients with damage to the ventromedial prefrontal cortex, who have impaired cognition of future time, make impaired decisions. Even when they have past experience of rewards and punishments for certain choices, they cannot make use of this "punishment schedule" when they make decisions for the future. Gerrans says they have a "myopia for the future;" they have lost the capacity that the rest of us have for "mental time travel." They have lost the ability to reason from the preferred future back to the present.

On the backward induction model, a decision maps out all the nodes in the sequence from start to finish. On this assumption, there is not much point starting a process if one does not know whether or how it will lead to a desired end. It is important for us that the forward and backward induction models, as well as other possible temporal models of decision-making, are likely to lead to different sets of decisions, and different results. Forward reasoning leaves an opening for the expectation that some alternatives not imaginable now may appear in the future and turn out to be the best choice—by leaving the future open, although, it confines the plannable elements of the decision to a zone of the future very close to the present. Backward reasoning, in contrast, second-guesses the long-term future to construct a more complete plan—by organizing the future, it makes a longer extended future into its zone of concern. This kind of dialectic is present in many of forms of decisional futures: Open future means less future; planned future means more future; but more planned future means less certain future.

In the big picture, forward versus backward decision-making—that is, creating values in the present and then imagining the future from the standpoint of the present, as opposed to creating values for the future and then imagining the present from the standpoint of the future—are not mutually exclusive. Decision-making imagination will always alternate among the two. Forward reasoning: If I make this decision, where will it go and what will I do when I hit the first obstacle (and then the next obstacle after that ... until I get to the end)? Backward reasoning: If I decide to aim for this goal, what obstacle will I have to overcome just before I get there (and then what will be the penultimate obstacle before that ... until I get back to the present)? Although back-and-forth speculations are likely to characterize every decision, it is possible that the two directions will lead to materially different choices. For example, if the decision whether to pursue distant space exploration focuses on backward induction from final goals, the number of unknowns may well suggest the futility of the project; whereas focusing on the launch-point as a first step may make the indefinitely long-term project sound more interesting. Any given decision might go one way or another. The point here is just that sequential decisions have complex bidirectional temporalities.

The cases we have dealt with so far involve a single person making a decision with a plurality of steps. Once we consider decisions in which many people have to anticipate what decisions other people are making, with all of their decision steps interacting, there will be a lot more complexity both in the reasoning leading to decisions and in their future reference.

The most peculiar scenario sometimes used to consider whether and how one person's decision should take another's perspective into account when deciding on the future, is the "Newcomb Problem." We will soon consider cases where decision-makers, without knowing what future decisions the others will make, need to take each other into account anyways. But the Newcomb Problem is a case where one person's decision hangs on another person knowing what the first person's decision will be (187–8). Newcomb's Problem: Assume that some amazing Predictor can predict with high probability what people choose. This is a ridiculous assumption, but assume it anyways. There are two boxes and you have to decide which to take: Assume that you know that Box A contains $1,000; and you know (don't ask how) that Box B *either* contains $1,000,000 *or* contains $0. You are allowed one of two possible choices: You can choose to have *both* Box A *and* Box B, or you can choose to have *only* Box B. It seems obvious that choosing both boxes is preferable. After all, if you choose both boxes, you will get the sum of what is in both, that is, you will either get $1,000 + $0 = $1,000, or you will get $1,000 + $1,000,000 = $1,001,000. If you choose only Box B, then you will get either $0 or $1,000,000. Obvious. But now make a second peculiar assumption: The Predictor will put the $1,000,000 in Box B, *if and only if* she predicts you will *only* take Box B. If she predicts that you will take both boxes, she will not put the $1,000,000 into Box B. Assume that you know that this is the Predictor's mode of operating. It now seems that, knowing this, it is rational for you to take only Box B, because if you choose both boxes, you will get $1,000 + $0 = $1,000, whereas if you choose only Box B, you will get $1,000,000. Now, the Predictor is making a prediction on the basis of some kind of evidence, and her prediction is presumably based precisely on knowing what you are thinking while you are making the decision. [Let us assume that this scenario does not allow the Predictor to look backwards in time, and let us not get into whether the Predictor is using magic—this part of the scenario is just not going to bear much analysis, so do not bother.] If you are thinking to yourself that taking both boxes is a rational way of hedging your bets, then probably the Predictor will predict that you will take both boxes, and then will *not* put the 1,000,000 into Box B. The Predictor needs to believe that it is likely that you will take only Box B, that is, the Predictor needs to believe that you believe that it is rational to take only Box B, if she is going to put the money into Box B and therefore make it rational for you to choose to take only Box B. The question is: Can you reasonably believe that it is best to choose Box B alone? With an emphasis on these points, we might conclude that it is. But now, there is another consideration, connected with causality and time.

The school of Decision Theory known as "Causal Decision Theory" adds a glitch. At the time you are making your decision, the Predictor is making her prediction. But by the time you actually take the Box, she has already either put the money into the box or has not. If you had deliberated about your decision by thinking that it was rational to choose only Box B, so that the Predictor predicted just that, and therefore put the money into Box B, the money is now, as a matter of brute fact, in Box B. Because it is now too late for the Predictor to switch out the money she has put in Box B (we stipulated that the Predictor does not move backwards in time), surely it is now rational to change your mind, and take both boxes after all. That way, you get the $1,000,000 in Box B, plus as an extra treat, the $1,000 that was in Box A. Of course, if you believed from the start that it would be rational to eventually take both boxes, the Predictor would have known that you believed this, and would have predicted that you would take both boxes, and therefore would not have put the money in Box B. And in that case, it will never have been rational to choose both boxes, as that choice leads you to miss the chance at the big money. The series of decisions only works out optimally if you do not merely pretend to believe (the Predictor predicts when you will pretend), but actually and rationally believe that you should choose Box B only; but then at the last minute, actually and rationally believe that it is rational to choose both boxes. In terms of rational consistency, this sounds like a contradiction. But it works differently once we focus on the fact that a change in the evidence has occurred once it is too late for the Predictor to take the money back out of Box B. In other words, switching seems possible because the situation at the later time is fixed in a way that the situation at the earlier time was not fixed. The fact that causality has a time arrow, makes it reasonable to make different rational decisions at different times. As Causal Decision Theory puts it, "the causal structure of the world is forward-looking, and completely insensitive to past events" (191). In one sense, flexibility in decisions gives decisions a future. But the implication of putting the first decision aside is that there is no point thinking that any given decision is the one that targets, let alone affects, the future.

The self-referential spin of the Newcomb Problem makes the decisional future a paradox. Only if you use what you know about the other's future decisions and actions can you fix the rational grounds for your own decision in the present (so you choose only Box B); but if you look too far into the future, to the point where the other has already acted on their decision, and try to get an advantage out of the fact that part of the future will have become past and no longer dependent on the other's decision (so you choose both boxes), then the other person will already have known that that is the way you think, and therefore would have subverted (by putting nothing into Box B) your tricky plan to act for the future at just the right moment in the time arrow.

There is no single solution to the Newcomb Problem in Decision Theory. Different theorists, using different formulas for playing the odds, and different ideas about how reasons and causes are related in decisions, and different ways of dividing the future into stages, solve it in their own terms. This complex and implausible scenario would be a long way to go if the conclusion were simply that decisions depend on what happens in the future. What is more interesting is the way decisions set goals and standards of judgment for the future. Those goals and standards themselves may be what subverts the future that one is trying for, especially if the future contains other people making decisions that likewise subvert your future, or if many people's future decisions subvert each others' goals and standards reciprocally. One might wish to simplify calculations so they will not to have to consider the other's interactive choices over time, but the obvious impossibility of doing so is why Decision Theory is inseparable from Game Theory.

The Prisoner's Dilemma

The paradigm game is of course the Prisoner's Dilemma. Two prisoners are being interrogated. If neither confesses, neither goes to jail. If both confess, both go to jail for a long time. If one confesses and the other does not, the one who confesses goes to jail for a short time and the one who does not confess goes to jail for a long time. The optimal result for both is if neither confesses. But if one suspects the other might confess to avoid a long jail term, then one might want to confess first, and be the one to get the shorter jail term. Again, there is no one solution to the Prisoners Dilemma in Decision Theory. Some theorists emphasize the optimal solution and see the dilemma as an argument for cooperation; others emphasize the need to avoid the worst, and see it as an argument for self-interest. The interesting thing is that the most rational self-interested decisions lead to suboptimal results: Both will confess, seeking an advantage over the other, and therefore both will go to jail for a long time. Even if the two self-interested prisoners communicate before deciding, and both agree not to confess, each one still sees an advantage if they renege on the agreement and confess first anyways. And assuming that self-interest is the driving force in their decisions, if they see an advantage in doing so, they will do so.

The lesson that Realpolitik strategists draw from this is that nations acting out of self-interest will want to begin a war before the other side does, because being peaceful in the face of the other nation's likely aggression will lead to its worst outcome. Both countries, then, will go to war at the first opportunity. The optimal situation for both countries would be peace, just as the optimal result for the two prisoners would result from trust. But motives of self-interest make that ideal mere wishful thinking. To avoid the natural result, our best realpolitikers used to say, the best we can do is make war so horrible that even self-interest will not lead to it. Nuclear weapons and mutually assured destruction is the only way to avoid the result otherwise predicted by Decision Theory.

As we will see, there is another way to avoid this result, but it depends on temporal infinity.

Again, contrary to popular presentations, Game Theory need not assume that self-interest is always the value that drives decisions. If some people value fidelity first, or if their hierarchy of values prefers the riskier best outcome over a more likely mid-level outcome, the prisoner's dilemma may be solved by neither party confessing.

The Prisoner's Dilemma is a "simultaneous strategy" game (220–22). The two players need not decide at exactly the same time, but each has to decide before knowing what the other decided. It is because of this that distrust leads each person to a decision which precludes the optimal result. Of course, simultaneity does not mean non-futural. The point is still to foresee what the other is planning, and to adjust your strategy accordingly. Decisions, in fact, are determined by a double future: The future with respect to outcome, and the future with respect to the decisions of the other player. And the other is anticipating the future of your decisions as well, so the second future is in turn a two-sided future.

The simplest case is where the players in a competitive decision all have the same evidence to work with, and all share the same rational process of decision-making. This assumption of "Consistent Alignment of Beliefs" presumes that "If you actually knew your opponents' plans, you would not want to change your beliefs" (Heap 25). The "Harsanyi Doctrine" presumes that "when two rational individuals have the same information, they must draw the same inferences" (25). The only rational ground for disagreement is if two people have different data sets (26), or if the one player does not know what the payoffs are for the other players (62). In the famous "Nash Equilibrium", there can exist a set of distinct rational strategies (one for each player), but it remains the case that the implementation of one player's strategy should be comprehensible to the other players, so that in the final result, each outcome, whether desired or not, "confirms the expectations of each player about the other's choice" (53). This does not quite require that all interactive decision-makers have the same beliefs, but it does require that each would not change their own beliefs if they knew the others' beliefs. The very idea of having the "same" belief is defined in terms of what one would not do differently the next time by knowing the other's belief.

Strategically, the flip side of this is that a player might deliberately "randomize" her behavior so as to be unpredictable, playing the Mad President to confuse the enemy. Thomas C. Schelling suggests that this "strategy with a random element" works well in competition games, but that in common interest games, there is a greater advantage in consistent behavior without surprises.[6] But I would like to imagine that there are games where freedom is the name of the game, where random elements aid cooperation, for example, in art-making games.

The assumption that players share rationality has a forward look: It anticipates consequences by assuming the other players will be rational in the future. It presupposes the principle of induction, namely that players will continue drawing inferences in the future the way they did in the past (98). When we talked of forward induction decisions earlier, we were referring to decisions based on the unknowability of the future, that is, decisions *not* to predecide what we might decide in the future. I would call that future-based forward induction. The case here, where a decision does predecide a future decision series based on the assumption that other decision-makers will decide in the future as they have decided in the past, I would call past-based future induction.

The tendency to self-interestedly be the first to confess has a different character in a "sequential" Prisoner's Dilemma, where the game takes place in several rounds. Before players make their decisions in a given round, they will have some (or all) information about the strategies used by the other player in earlier rounds (Peterson 222). This way, players can learn about each other and form new strategies round by round. They might only know which choices the other player made, or they might have access to all their deliberations. Over time, they may see how the other player learns new strategies over time. This is an area where Decision Theory could learn from phenomenological categories of intersubjectivity and temporality. The complexity of mutual, nonindependent, thinking processes does not come into Decision Theory much, and this is in good part why the whole field seems naïve to phenomenologists. Both fields would benefit from bringing sequentiality and intersubjectivity into the same context.

In an "iterated" or "repeated" game, the entire game is repeated a certain number of times (224). Imagine a Prisoner's Dilemma played out each day, with gains and losses distributed each evening, and a new game begun the next morning. In most types of sequential games, a "tit for tat" strategy might be rational. Each player can decide to confess or remain silent, based on what the opponent did on the last round in the game, or the last time the game was played. If you remained silent in Round 1, and your opponent confessed, you can get even in the following round. If your opponent remained silent in Round 1, you can take that as a treaty offer, and remain silent in Round 2. It is true that the other player can then renege in Round 2, but then you can take revenge by reneging in Round 3. If there are enough rounds to stabilize expectations and create a momentum of agreement, then you can start to assume that the other player will likely remain consistent with their previous decisions, which function like a promise. In these conditions, even the most self-interested prisoners might play a cooperative strategy. Each player can risk losing out on a given round, in order to communicate to the other her willingness to cooperate in the following round. The future only allows for self-interest to transform into cooperation, and for suboptimal scenarios to be replaced by optimal results, if there is willingness to sacrifice temporarily. Without sacrifice, the future repeats disaster.

If players could communicate directly, and talk about what they planned to do, it might seem they would agree not to confess. But this is not enough. The problem remains that if either one betrays the other and confesses, while the other honorably does not, the traitor will get a better result than he would get by not confessing. Therefore, once again, it is in the interest of each to make the promise and then renege on it. "It is very difficult to make people believe your intentions when you have an incentive to lie" (Heap 117). So far, the future is still determined less by communication than by interest.

In fact, there is a "problem of trust in every elemental economic exchange because it is rare for the delivery of a good to be perfectly synchronized with the payment" (149). This is an important point: It is the unsynchronized timing of payoffs that leads to distrust, and makes intersubjective time problematic in decisions.

In many decisions, we would like to do the right thing as long as other people do it too (conserving electricity by keeping our houses cool, paying our taxes, joining the union, etc.). But there is the "free rider problem" (151). Experiments have been made with multistep role playing games, where participants are assigned roles defined by inequities of value, for example, by gender or race. We might assume that it would be to the advantage of underprivileged members of the society to rebel, particularly in "evolutionary games," where decision-makers "adjust their behavior on a trial and error basis toward the action which yields the highest payoff" (195). But, if we believe the experiments (probably we should be skeptical), players rebel only if a large enough number of other people rebel also. If others do not rebel, the inequities become stable (213), the losers become marginalized, and thereby become less likely to communicate with each other or even to know that the inequities are becoming more and more systematic (214). It seems on this model that a self-interested person would like to get the benefits of communal action without taking the risk of it, to get the good end result of a process without going through the period of sacrifice that it normally needs at the beginning, that is, that a self-interested person would freeload on other people's risk-taking. Of course, if everybody decides to be a freeloader, the optimal result will never be obtained for anyone. As in the Prisoner's Dilemma, cooperation improves outcomes, but it only works if almost everyone takes on the role of cooperator. To decide whether cooperation is likely to emerge around you, and whether you will have to contribute personally, you have to estimate the probability that the other players are cooperator types. A given decision-maker might of course choose a cooperating strategy even if he is not the type (181), or vice versa. A person might decide to be an all-the-time cooperator, or a "constrained maximizer" (162), or play a "tit for tat" strategy (165).

In exceptional cases, features extraneous to a game may create cooperation even without communication (205). For example, when people are told to meet in New York, but not told where or when, most will go to Grand Central Station at noon. If told to pick heads or tails, most pick heads. Prominent facts, social context, and aesthetic norms can influence decisions where no rational grounds exist. A certain amount of cooperative decision-making is simply a bandwagon effect (208).

Normally, though, cooperation takes time to develop. It takes a repetition game, where each player sees what the other player does a few times, that is, where each player builds a "reputation," before cooperation becomes the norm. As we have seen, running the game just a few times still leaves betrayal as the option that looks most likely to lead to one's best result. It is true that a player does not want to be punished by his opponent in the next round for his betrayal, but this just means that he will wait until the last round to stage his betrayal. Players will cooperate as long as the game is ongoing, but both will renege at the end of the sequential game. This is a sorry state of affairs, as they know that their final outcome would still be better if both did not renege (Peterson 252). But what can you do? The twisted state of the player's mind is after all how Decision Theory helps us to predict decisions made by deceptive people, like to predict whether Congress will decide to cooperate with a President (Heap 187–9).

But what if there is no end to the game sequence? Or more modestly, what if the players do not know how many times the game will repeat? In "infinitely repeated games," there is no round such that one cannot be punished on the next round. Therefore, there is no round on which betrayal carries no risk. Therefore, there is no round on which one player knows the other will betray him. Therefore, both players can cooperate continuously without fear of immanent betrayal.

To yield cooperation, the game really has to be, or to be thought to be, infinitely repeated. A game with a very large number of iterations is not good enough. As soon as the players know there will *eventually* be a last round, then they both know that they will both confess on the last round. But knowing this, they know that if they renege and confess on the second to last round, there will be no time for their opponent to punish them in the final round, when they intend to renege again. But then, each player might as well renege on the third to last round, and so on for all the rounds. They might as well renege on the very first round. Variations on this paradox are known as Centipede Games,[7] presumably because the first of a hundred legs moves the leg a hundred joints back.

In a way, it is more important that players think that the game is infinite than that it actually be so. A particular game might in fact come to an end with a last move, but if neither player knows when or whether that will happen, decisions are made as if it were an infinitely repeated game. As far as decisions are concerned, what makes a future open-ended is the virtual future, not the empirical future. The virtual future makes the decision-maker free to take a chance, whereas the empirical future binds the decision to self-interest. Although Decision Theory would not put it this way, the future in Decision Theory is phenomenological; it exists in thought. The outcome is measured quantitatively, but the future is intended qualitatively.

The point is that without the certainty that the other is going to betray them in the final round, both players can rationally postpone their betrayal forever. The motivation in finite games for moving the betrayal back retroactively from the final round to the first, does not hold. There is no backwards induction from an infinitely deferred decision (Peterson 253). There is no moving backwards from a final result if there is no final result. There is no motive for betrayal if there is no Aristotelian final telos of decisions.

By mutual postponement of betrayal, both players can achieve the optimal results that distrust had been keeping them from. Self-interest still makes a player lean toward betrayal, but the fact that the game keeps repeating, allows players to change their mind every single time, and convert from betrayer to cooperator over and over—each time just temporarily, but repeatedly. The temporary choice becomes the choice repeated over an infinite amount of time, to the benefit of both players. The finitely iterated game, the game with a merely finite future, with a last round no matter how long delayed, will eventually replicate the poor result of the one-off, non-iterated game. Only the infinite future allows for repeatedly changing one's mind in the optimal direction in the future.

Of course, deliberations will still be complicated. A large number of repetitions can create a large number of equilibrium strategies—different pairings of strategies that will maximize results for one member based on what the other chooses. But generally, more cooperation, and therefore better results, are achieved when players focus on an infinite number of steps before and after the local decision node. They decide best when they are forced into hesitation by not being able to picture a final iteration. Combining these two issues—sequence and intersubjectivity—produce patterns of decision formation ranging from 12-step support groups to perpetual revolution. One might wish to pick a "dominant" strategy that works no matter what one's opponent does. But in games with unending repetitions, all sorts of pairs can bring best results, so there is no best way for two players to play the game against each other. Variability is the real topic of theory.

Now, the other's offer to cooperate still only works if each player "knows that the other knows that ... the other knows that ... etc." (225), that is, if all players anticipate how the others think. The only way for there to be an open future, and thus a chance for optimal outcomes, is if the future is intersubjective. But intersubjectivity need not mean symmetry. Suppose your opponent betrays you in one round by confessing? Does it follow necessarily that you should confess in the next round? It might, if decision-making were modeled by algorithm, but decisions are made by imperfect humans. Sometimes a human being just slips up and pushes the "confess" button by accident. Any serious decision-making calculus needs to take accidents into account. It sounds funny, but it is natural that Decision Theory builds in a "Trembling Hand Hypothesis" (254). If you think the other person just made a mistake, and meant to make a different choice, it is rational to act as if it did not happen, and therefore to break the round-to-round symmetry.

In a "perturbed" game, there are small "trembles" of this nature, which players can take into account, possibly changing their strategy, without knowing their exact probability (Heap 68). A tremble can be assigned a small theoretical probability even when it is hard to believe that someone could make such a mistake. A player might even deliberately build small errors into their strategy (a type of "mixed strategy), or "bluff" in order to appear irrational (90), to confuse the opponent (71) (as a blackjack player might make small errors so the casino will not know he is a professional). For our purposes, we might call these not just tactical trembles, but temporal trembles.

Indeed, even hard-core competitors, who aim to win by means of threat, recognize that the game is played by what does not take place at least as much as by what does. As Thomas C. Schelling, an important figure in early Game Theory, says, a threat is only what you will "probably" do (Schelling 14). A threat strategy does not require applying maximum force, it only requires "exploitation of potential force," getting maximum results from the fact that the other players know your potential. We might even say that Decision Theory employs the principle of minimum action. "A 'successful employees' strike is not one that destroys the employer financially, it may be one that never takes place. Something similar can be true of war" (6). The whole idea of deterrence is to win a war that never takes place.

Once we know that there are all sorts of stops and starts, miscommunications, lures and trembles, repetitions and iterations, then we can describe decisions as structural and temporal matrices. "Dynamic games," or "extensive games," can be diagrammed to show when a series of choices are to be made, and when they intersect at "nodes," "chapters," or "subgames" (Heap 82).

One of the most interesting consequences of dynamic sequential games is that "a player might not know exactly where he or she is in the tree diagram" (83). For that matter, it is not even clear how we know when a decision is called for, or which moment counts as a decisive turning point. This is the problem of "decision point recognition." It is possible that some kind of "somatic markers act as an 'alarm bell' in redirecting attention," although it is not easy to locate markers in infinite sequences.[8]

A decision-maker may not know which of several similar moments in a sequence she is located at. She may not know whether the other player has already moved, or how many times, or whether something in the state has changed since the last time she moved. If she is not sure which step of a decision she is making, she has to contemplate the decision as if she is at a number of different time lines. She may find it undecidable which time line is actual, which are only apparent, and which are potential. The decision of which stage of the game she is at may be its own node in the game itself (93), or it may be more like a meta-decision. The more stages there are in a game, the more complexity this raises. Obviously, in practical as well as theoretical terms, it is crucial for all decisions that the decision-maker know at least largely what has happened so far, that she know what stage in the process she is intervening in, that she know whether the process is just starting or coming to an end. For example, when deciding on a large scale how many resources should be put into saving future generations from global warming, it might be relevant to judge whether the human species is just starting to evolve or is nearing its end. And when deciding on a small scale whether to stop for gas before the end of a drive, it is essential to know how close one is to the end of the trip. If a batter forgets how many men are out, if a pension plan purchaser misjudges his life expectancy, if a soldier cannot figure out which command applies to her and which applies to somebody else, if a musician performing a repetitive piece by Steve Reich has not counted how many repetitions have been completed so far, if a prisoner can create confusion around his ID number, then the patterns of seriality can be messed up in a way that damages, or improves, expected outcomes. All large scale decisions admit some confusion about which node in the sequence is being carried out by a given decision. The more difficult it is to determine at which point in the overall process one is positioned at when making a particular sub-decision, the less one can tell whether induction toward a desired goal is forward or backward.

Uncertainty and temporality

There is almost always a degree of uncertainty about which decision will lead to the best outcome. This is central to the idea that the future is virtual.

Fuzziness

To deal with the fact that so many decisions are made under uncertainty, a subfield of decision theory has developed around fuzzy logic.[9] A set is fuzzy if some elements neither belong nor fail to belong to the set definitively ("crisply") in an either-or manner, but belong to it to a degree, depending on relevance and context. For example, a person may belong to the set of "young people" to a degree, depending on whether the set is formed to pertain to education funding, to marriageability, or to some other decision-worthy issue. One advantage of formulating Decision Theory for fuzzy sets is that many decision-makers, whether they are suckers deciding to buy a used car, or corporations deciding to build a new factory, find they have incomplete or ambiguous information, then despair of being able to calculate a good outcome, and so resort to intuition or some other ridiculous way of deciding what to do.

If people could get familiar with fuzzy logic, the theory goes, they could be rational in spite of incomplete information. A second advantage to being able to reason under fuzziness, is that gathering more information sometimes has too high a cost. If decision-makers feel the need for crisp information, they may waste a lot of money on clarifications that have little utility (Rommelfanger 5–6). On the downside, fuzzy logic decisions may not always generate a clear winner among decision options. But on second thought, that is also an advantage, since in the real world, there sometimes simply is no best choice (19).

There are various fuzzy stages in decision-making. When there is fuzzy information about the world (e.g., about how many hungry people one would have to feed if one were to decide to fight against hunger), there is a "fuzzy database." Where there are variable switches that change the situation (e.g., whether to fly through an airport that often closes for bad weather), there is a "fuzzy decision table." Where community values are disputed, there are fuzzy parameters. The goal is to know precisely how fuzzy a set is, and to assign outcomes "fuzzy probabilities." Most important for us are the situations where a decision has "fuzzy consequences," or "fuzzy expected values." For us, this opens the problem of measuring time on a fuzzy future.

Of course, from our perspective, a mathematical approach to the fuzzy future repeats the problem it started with. To assign an exact degree of fuzzy probability to a set of objects, to a changing background condition, or to a shared value, limits the possible ambiguities that can arise over time when events hit snags and accelerations, and have to be measured in the future by parameters we do not yet foresee. It presupposes that each benefit parameter (e.g., the price of buying a car, the environmental risk, the safety, the pleasure factor, etc.), is clearly defined, even if the value of the car under each criterion is fuzzy.

Decision Theory is certainly not limited to decisions where there is only one criterion for valuing outcomes. A sub-area of research deals with "Multiple-Criteria Decision Analysis," to calculate preferred options under a combination of inconsistent values. But here again, the objective of this sort of analysis is to package fuzziness and multiplicity into a single unambiguous formula, leading to a single preferred decision, aiming at a single temporal series of singular steps with a singular final result.

Would it be possible to capture the fuzziness of the future itself? It would not be too hard to describe as a fuzzy set the different amounts of time, or the different number of sub-decisions, that it might take to carry out a decision. But might we say in addition that the designation of which temporal moment comes "next" after a given moment should be described as a fuzzy set consisting of several moments, even of which belongs to the set with a different degree of intensity? Could many different times be "next" after a given time by degree, so that a time generally much later than time-1 might nevertheless be proximate to time-1 to a small degree? Could several different times have the same small degree of "nextness" after a given time? Could we define virtual times as times that have a small but non-zero degree of presence in the actual time line? None of these formulations sound ideal, as they all assign numerical values to probabilistic time relations, reducing virtuality either to degrees of predication or to degrees of probable instantiation. Still, fuzzy outcome and multi-criteria analysis are ways in which Decision Theory reaches for a phenomenology of decisional futures.

Reversibility

Under conditions of uncertainty, it is generally preferable to make decisions that maximize the room for future revisions. In my terms, decisions that allow for, call for, and provide resources for more future decisions are both more decisional as noetic acts and more futural in their noematic reference. Existentialists and postmodernists can agree with this: Free decisions are decisions for freedom. In Decision Theory, there is a way to articulate the added value of reversible decisions, decisions that do not block the option of making different decisions later.

On Claude Henry's[10] definition, "A decision is considered irreversible if it significantly reduces for a long time the variety of choices that would be possible in the future" (Henry 1006). For example, the decision to demolish Notre Dame cathedral would be irreversible, but the decision to preserve it is reversible.

Henry's research focused on the planning stages prior to the building of the Periphérique highway around Paris. Cost–benefit planners calculated values to be gained by building the highway, such as values gained in trucking time saved, against the values that would be lost, such as loss of public forests and ancient royal estates. Bureaucrats tend to make irreversible decisions (1007), but, he argues, reversibility itself has a significant value. All sorts of things can go wrong while building a highway: The ground soil in some areas could prove problematic once the digging starts, or an archaeological site might be discovered along the route. To prepare for such eventualities, many different plans for the route should be worked out in advance, with contingency plans to switch parts of the route in midstream. Until each segment of the route is ready for construction, no buildings on that segment should be demolished, just in case that piece of the route has to be replaced. A plan should leave as many of its elements reversible as possible.

To calculate the value of reversibility, we would have to measure the number of stages in a decision series, and to know at the end of each stage how many options remain for the next stage. If A and B are options, and a decision chooses A, but a subsequent decision can switch it back to B, then the decision for A is reversible. If, in addition, B, C, and D are available as independent options while the A-or-B decision is being made, and B, C, and D are all still available after the decision to choose A over B has been made, then the decision for A was fully reversible.

For our purposes, it is not crucial whether the future always leads to better information (as Henry thinks), or even whether all steps of all decisions are temporally locatable; it only matters that the decision's motives, information base, and/or values can change in the future. It is enough to point out that a decision-maker will want to control the next decision as well as the present one.

There is a still more complex situation for "decisions which would be made now and applied after a delay" (1007n). By the time a decision is applied, the very fact that the situation was made and announced earlier may already have caused there to be more or fewer options. In my vocabulary, it is as if the decision is being made at two times, the second of which has already been affected by the first. It is as if we make a decision after having seen some of its future results. If making the decision has already reduced its own options, it may already have excluded the option of carrying out that very decision.

It is complicated to describe mathematically how great the reversibility effect is, that is, what weight should be assigned to the loss of potential that comes from an irreversible decision. This may depend on how long it takes for the returns of a decision to become clear. If the return on a decision is more or less immediate, the impact of irreversibility is low; if the return is far off in the future, the impact of irreversibility is likely high. Henry calculates that in a decision where the effects' deferral into the future is "moderate," choosing a reversible option over an irreversible one gives an added value of thirteen percent (1011). I am not qualified to check his calculation.

Of course, reversible decisions accept a higher degree of decision anxiety, but anxiety is the price to pay for a reason. No doubt, many people would like to eliminate uncertainty before making the decision. But George Wu finds that "in many real-world gambles, a nontrivial amount of time passes before the uncertainty is resolved but after a choice is made."[11] If uncertainty-averse people are given an option between a course of action that will show results quickly for better or worse, and a potentially better option where the time to resolution is unknown (a "temporal lottery"), they prefer the former. Pre-tenure tenure-track assistant professors at a good school may thus prefer immediate tenure at a lesser school. Only rarely do people prefer uncertainty (as when they buy lottery tickets). Wu argues that it would be rational to accept longer periods of delayed resolution than people usually do. In my terms, it is rational to accept longer post-decision unresolved futures as part of the decisional noema.

Improvisation

There are extensive empirical studies of improvisation in areas as diverse as jazz performance, business operation, and emergency management. The Center for Technology and Behavioral Health at Dartmouth College's project on "Flexible decision making in response to unplanned events"[12] studies "improvisational" decisions with "momentary disruptions and plan deviations" in the field of drug abuse prevention. Practical studies of "real-time dynamic decision making" carried out "on the move"[13] raise issues both of teamwork and extemporaneity, or "making do." Decisions of these sorts cannot be fully planned, either by intersubjective preparation or by stockpiling resources. Nevertheless, businesses train for and reward improvisation.

Of course, improvisation grows with detailed technical knowledge, rehearsal, and anticipation of responses. Improvisers in jazz study harmonic progressions and substitutions, memorize and vary phrases and idioms, practice interplay between players, and submit to formal constraints to release degrees of freedom.[14]

Polyrhythm

I mention just one case of polyrhythm, drawn from hotel management. The main temporal division in hotel decision-making is between the "front of house" (the service area) and the "back of house" (the administration area, not to be confused with the "back area" [the kitchen]).[15] The two zones have different hours of operation, different speeds of activity, different types of deadlines, and in general, different rhythms. There need not be "complete synchronization of each event with all the others ... 'It is enough to be able to stage in each moment what is simultaneous with what,'" even though they "run on different clocks."[16]

Uncertainty and systems theory

Systems Theory is in some ways a version of Decision Theory, pertaining to decisions about complex systems with interacting parameters and feedback loops. But it also poses a challenge for Decision Theory, in that such systems can be so complex that interventions tend to have unintended effects. Even well-meaning decisions tend to be either useless or destructive, and the best thing to do is usually to leave the system alone, until it reaches its own equilibrium state.

Jay Forrester's[17] applications of feedback loops to urban social dynamics in the 1960s reveal both the strengths and weaknesses of systems analysis for a theory of decision. "'Feedback loop' is the technical term describing the environment around any decision point in a system. The decision leads to a course or action that changes the state of the surrounding system and gives rise to new information on which future decisions are based" (Forrester 107, 13). Many human situations are not systematic in this sense; they involve linear processes and only show the effects of negative feedback. For example, when we go to pick up an object from a table, we get our hand near to it, then notice a difference between where our hand is and where the object is (negative feedback), then move our hand a bit more, and so on, until the goal is reached. This is obviously not a great phenomenological description, but it may be true that in such a case, there is only one parameter for success, one cause–effect relation, and one temporal sequence of before–after positions. In complex situations, however, where there are dozens or hundreds of interacting parameters, as in urban societies, changes are nonlinear. A change in one parameter can have a vast set of effects on other parameters before it has its effect on the parameter it began in. As it is only through those indirect interactions that the effect takes place at all, it is effectively impossible to establish a linear trace from an effect back to a single cause, or vice versa (9).

Nonlinear systems with feedback loops have been studied in industrial, urban, and global social dynamics, as well as in computer networks and natural ecologies. To use one of Forrester's central examples: Well-meaning 1960s social reformers frequently decided to increase the minimum wage in a given city, with the obviously desirable goal of decreasing poverty. However, on increasing the minimum wage in a given city, it could well happen that more unskilled workers will move to that city, which will not only create inner city crowding and worsening housing conditions, and thereby add to the tax burden of the middle class, who will then move away from the city, and thereby reduce the number of enterprises in which the unskilled workers might otherwise have been able to get jobs, it will also create a large number of competitors for the minimum wage jobs, and thereby reduce future pressure for the employers to increase wages. An intervention intended to raise wages in a complex system could thus easily have precisely the effect of suppressing wages, once the initial intervention has worked its way through the system. For Forrester, unintended consequences like these are the norm for decisions targeting nonlinear systems. In the short term, the minimum wage will of course increase, but over the long run, the increase will not be significantly higher, and may even end up lower, than if the wage had increased very gradually on its own. In highly complex systems, generally, interventions that introduce a movement into the system tend to create countermovements in the system, and in the end, they either have little effect, or they produce counterintuitive effects opposite to those intended. "Nonlinear behavior makes the system resistant to efforts to change its behavior" (108).

Indeed, the situation is even more "devious." Since the first intervention does have a short term desired effect, this appears as "positive feedback," and decision-makers will repeat those apparent successes, increasing the minimum wage again and again. But the more they intervene, the greater and more disastrous the counter-effect will be in the long run. What makes the decline difficult to see and take into account, is that there is typically as much as a twenty-year "perception delay" before city planners notice a problem. And even at the point when they notice that the minimum wage increase has made low-cost housing more crowded, and that new housing has to be built, which is now more expensive, those planners will think that the best way to solve these new problems is to raise the minimum wage again, thereby bringing the situation closer to failure with each iteration of the apparently successful strategy.

Anti-interventionist views like Forrester's in the 1960s generally meant opposition to liberal reforms in the war on poverty and civil rights. But the same anti-interventionist attitude should be critical of any interventionist policy, including more recent conservative directions like the war on drugs, war on terror, war on taxes, or the idea of building a wall to keep immigrants out of a country.

One interesting implication of feedback loops is that it does not much matter how large a system is when it begins (38)—cities from one million to five million will exhibit the same kinds of system effects. Of course, early in the simulation run, the cities will behave differently, but structurally similar systems with different initial conditions will converge to a similar pattern in the long run. This is rather surprising, but it is confirmed by the fact (at least, Forrester takes it as a fact) that cities of all nations, all time periods, and all levels of access to resources, undergo the same cycles of growth, crisis, renewal, and eventual equilibrium, over a time period on the order of 250 years. And it is not just size that is largely irrelevant to the ability of a system to absorb interventions without their histories being changed. Differences of culture, of affluence, and of geography are all absorbed by the system. "Multi-loop realignment along various nonlinear functions makes the complex system highly insensitive to most system parameters" (108). Methodologically, this means that system theorists ought to put their effort into designing generalized computer models to simulate system dynamics, and not put much effort into empirical histories of a given society, as those factual coefficients are boiled out of the system over the long run. And city planners should allow urban systems to propagate over time, rather than introduce special effects into their own cities to make them diverge from the typical pattern.

In one sense, computer simulations could be regarded as a cybernetic way of predicting and controlling the future. However it would be more correct to say that the future in the usual sense, where something unexpected might happen or some new expectation might arise and be acted on, is rendered obsolete. For Systems Theory, uncertainty pertains to the effects of intervention, not to systems left on their own. In other words, "this uncertainty about future dynamic implications of assumptions in a model is totally eliminated in computer models."[18]

It is not that decisions play no role in systems application. On the contrary, decisions that ignore a system's tendency to equilibrium can play a very unfortunate role, namely by wasting resources, or preventing the system from bearing fruit naturally. Minimal decision-making reaps maximum benefits. Still, the idea of decisions fulfilling human projects by making the world-path diverge from how it was going, is, for Systems Theory, an illusion.

Of course, Systems Theory might just be mistaken. Some recent studies appear to show that in many systems, from natural environments to political societies, there is no tendency toward equilibrium after all.[19] And if there is no eventual equilibrium in a system, then feedback loops do not necessarily entail that interventions are ineffectual or counterintuitive. Of course, even if systems do not tend to equilibrium, they will still be nonlinear. There will still not be linear series from a single cause to a single effect along the same parameter, so decisions will still need layered and branching analysis. But at least decisions may be more decisive than the equilibrium version of Systems Theory predicts. (I do not have the expertise to judge whether current evidence supports or undermines assumptions of eventual equilibrium.)

But there is another assumption of Systems Theory that seems incorrect, namely its assumption that systems are bounded.

Forrester assumes, for example, that a given city grows, shrinks, renews itself, and reaches equilibrium relatively independent of the rural, national, international, and planetary environments on its outside. His computer simulations do take into account population influx and outflow, biomass consumption and waste emission, pollution, and interstate transportation. However, he figures that once such patterns are known, urban planners can treat their cities as relative totalities. I think it is fair to say that this assumption is exaggerated. At any rate, for those systems with open boundaries, whether we are talking about an individual's life, or the creation of a work of art, or a political revolution, or space travel, or the world population's will to change its energy source, decisions will not necessarily be swallowed up by a system's internal tendency toward equilibrium. So, although we can learn from Systems Theory how to use nonlinear feedback loops, we will have to look elsewhere if we want to find steering procedures for unbounded systems.

Finance

In the 1930s, the "Theory of the Firm" introduced rational decision-making into corporate finance theory as a way of predicting how firms interact with the market. In the 1950s and 1960s, this was developed into "Efficient Market Theory," to explain the rationality of market fluctuations.

Finance prediction is complicated by the fact that the two elements in markets—manufacturers and the stock market—are in some ways independent. The financial advisor Daniel Peris[20] has a good explanation of the distinction between long-term "investment" in a company, which pays off largely in modest but relatively predictable dividends if the company makes and sells a good product, and short-term "speculation" on stock market prices, which may fluctuate independent of the company's product. Speculators rely on less steady factors, and in that sense their decisions seem less quantitative; but on the other hand, they care only about short-term stock prices (investment "performance") and not about product quality, and in that sense, their decisions seem more limited to a single quantitative metric. Focus on stock prices has the advantage of holding company managers accountable to investors, but the disadvantage is that annual, quarterly, or even daily fluctuations of stock prices override longer strategy, and give incentive to optimistic or even fraudulent earnings reporting.

Nowadays, the school of "Behavioral Finance" challenges market rationalism, arguing that emotional factors such as irrational degrees of risk tolerance or risk aversion affect the market as much or more than any rational decision procedure. Daniel Kahneman's[21] "Prospect Theory" finds that decision-makers most often use heuristics instead of thorough analysis, shorthand versions of logic and statistics that may lead them to decent decisions in most cases, but to disastrously bad decisions when the shorthand versions miss crucial details. His thesis, which seems hard to doubt, is that people generally do not make decisions consistent with "Expected Utility Theory." If he is right, then on the one hand, decision-makers cannot be relied upon to make rational judgments; but on the other hand, their errors will be predictable.

There is an old idea, not universally accepted, dating back to the nineteenth century, called the "Random Walk Hypothesis," which holds that independent stocks on the market fluctuate unpredictably within a relatively limited range. Like dice, individual results need not be representative of statistical patterns. The usual interpretation is that the random walk is consistent with market efficiency. The idea of market efficiency is that the market always already prices stocks within an appropriate range, and as long as random fluctuations are held within that range, they are statistically insignificant, and therefore unpredictable, but also not destructive, in the large picture. The assumption is that information regarding a given stock is already embedded in its price. A small investor is never going to know more about the stock than the big investors have already known; the big investors have already used this knowledge to buy and sell, and therefore to create the price of that stock. It is absurd for a single small investor to believe that he or she knows more than the market knew a while ago, and it is therefore foolish to think that he or she could predict when a stock will go up or down. It has already gone up or down, and all that is left are random fluctuations.

Before the Efficient Market Theory, investment theorists had the idea that stock prices on the market are often over- or undervalued, and that a rational investor could find those errors and make money by anticipating corrections. But once the Efficient Market Theory took hold, theorists concluded that because a small investor cannot make money by special insight, their best bet is to invest in the overall market with Index Funds, and wait for the market to climb. With an Index Fund, there is nothing to predict, no decisions to make, no future noema, just a continuing advance of the present. Nowadays, when the Efficient Market Theory itself is in decline, and many theorists think that investor behavior is marked by the same irrational patterns that mark all human cognitive judgement, there is again the hope that a small investor might find lucky bargains on the market.

Long term observation obviously displays crises in the history of the stock market, and major bubbles and busts, and these obviously have a huge impact on society, not to mention on investor results. The Random Walk Hypothesis does not explain these. The risk analyst and statistician Nassim Nicholas Taleb argues[22] that random exceptions, although rare "black swan" events by definition, are so powerful in creating crises in the market and elsewhere, that the most "robust" investment idea is to hedge one's decisional bets, to predict only that the market will be unpredictable, to invest in products and services that will be popular in times of crisis, so as to make money when fragile things inevitably fall apart. No doubt there are exaggerations in Taleb's view of social fragility, but there is something akin here to the promotion of reversible decisions that attend to the futural character of events.

It is not clear that we should build an account of future directedness on any of these market analyses, as among other things, they all ignore the future of labor in the economy. The theories I have mentioned seem interested primarily in whether prices are right. But, to take one finance-related example, arbitrage is all about putting companies into bankruptcy to cancel union contracts, reduce or export workforce, wages, benefits, and pensions. To articulate such practices in terms of a company's value–price ratio, while ignoring the social and political meaning of investment, is to evaluate quantities instead of anticipating futures. The paradox is that investors who design future profits avoid contemplating the future of the real world.

At any rate, the skeptical conclusion is that the more data we have, the more we see how little can be predicted. Skepticism about predicting financial markets goes back at least to John Maynard Keynes in the 1930s.[23] Keynes' judgment is that we overconfidently expect the future to be like the present (Keynes 97), a "convention" that makes investments seem safer than they really are (100). In fact, "our knowledge of the factors which will govern the yield of an investment some years hence is usually very slight and often negligible" (98).

One might think that instead of predictive mathematics, we could turn to risk management, to minimize the risk that comes from not knowing what we should decide. But risk management still claims to predict the mathematical relationship between risk and return. The risk management offices of financial institutions were among the worst culprits during the financial crisis of 2008.

If we wanted to work out a complicated case, like how the financial crisis of 2008 arose, the issues appear so dauntingly complicated that it is difficult to be sure what is predictable and what is not, and how we can think about the financial future in anything like a plausible way. Looking backwards, we can see how the Clinton era instruction to Fannie Mae (the Federal National Mortgage Association) to lower interest rates so people with less wealth could afford houses, along with the Federal Reserve's interest rate policies, interacted later on with private providers offering subprime teaser mortgages with low starting interest rates to poor people knowing they would have no ability to pay once the rates went up three years later, with optimistic assumptions about perpetually rising house prices despite stagnant wages, with the new practice of mortgage writers selling new mortgages to banks and financial institutions, with the new derivative risk distribution mechanisms where small fractions of thousands of different mortgages with different degrees of risk were packaged into poorly understood financial products like mortgage-based CDO (Collateralized Debt Obligation) bonds purchased by large institutions for amounts in the hundred million dollar range, along with the new instrument of Credit Default Swaps that allowed large speculators to bet on the failure of other large speculators, the strange invention that multiplied by several powers the losses that would accrue when any large group of investments went bad (without this last step, a mortgage crisis might have meant a trillion dollar loss, but that could have been absorbed by the large American and international economy; it was the secondary betting on that trillion that led to the unabsorbable multi-trillion dollar losses), along with the financial institutions' poor assessment of risk and overcommitment to speculation, made possible by the weakening of the division (previously regulated by the Glass-Steagall Act of 1932 to prevent Depression-era crises from happening again) between the commercial and investment branches of financial institutions, along with the quasi-monopoly and gamability of the ratings agencies that understated the risk of the mortgage bonds, along with the Basel accords that decreased the amount of liquid assets that banks have to have on hand in case of investment failures, which worsened the stressed banks' liquidity crisis of overnight short-term paper, of course along with decisions made after the crisis began by the George W. Bush cabinet about which firms to save and which to let fail, and TARP's (the Troubled Asset Relief Program, whose council members Bush and Obama chose from among the firms that caused the crisis) distribution of $620,000,000,000 again to the same firms that caused the crisis. In a situation like this, it is unclear who makes decisions and what future is projected with any determinacy at which point along the way.

But even short of this sort of overdetermined system, the possibility of prediction-based financial decision-making is already full of difficulties.[24]

To study human prediction empirically, the political scientist Philip Tetlock established the "Good Judgment Project."[25] Thousands of people from all walks of life volunteer to undertake all sorts of predictions about global events. Each predictor can choose to use any data from any source. They can make as many predictions as they like and update them as often as they choose. Tetlock checks their results, then analyzes what makes the best of them, the "superforcasters," better than the others. Of course, successful forecasters have to understand base rates and principles of statistics and probability theory, as well as the special statistical properties of rare events. But for our purposes, Tetlock's study has two other main findings. First, forecasters who make frequent small updates in their predictions are by far the most successful (Tetlock 154). Second, forecasts mean nothing unless they have a specific time reference (52). It means nothing to predict that there will be a downturn in the economy, unless you say when. Both noetically and noematically, predictions improve as their future structures become more fine-grained. As an alternative to skepticism about market predictions, we might expect theorists to work on subtle interactions between short term and long term. Instead of uncertainty leading to a purely indeterminate future, uncertainty might lead to peculiarly complex futural structures.

Oversimplifications arise when we expect the future in the long run to be like the past, but only apply this insight in the short term. Or perhaps we credit it in the long term but with such undeserved certainty that we expect it will be confirmed with each short-term time segment. Manipulating investor assumptions, and of course their greed, securities markets encourage people to revalue their investments several times a day—too often to be rational, as it leads to buying and selling stocks too soon (Keynes 99). The upshot is peculiar. The probability of long-term predictions are low, but the probability of short-term predictions are lower. So it may be better after all to form some long-term expectations than to commit all one's resources to the short term. Keynes does not put it this way, but improbable long-term decisions seem at least as likely to pay off as probable short-term decisions. There is no doubt a kind of advantage in short-term decisions, in that they value liquidity, flexibility, and reversibility. But in Keynes' view at least, they are "anti-social." They give up on the shared intersubjective future. "The social object of skilled investment should be to defeat the dark forces of time and ignorance which envelop our future" (102). If people really did not trust either the short term or the long term, that is, if they knew how unlikely it is that their investments will yield big profits, most people would not invest at all, and that would be a problem for the economy. It is the largely irrational "animal spirits" of optimism, and the projection into decisional futures, that, fortunately for the economy, generate faith and hope in investment, in the same way that "a healthy man puts aside the expectation of death" (106).

Keynes' conclusion could virtually serve as a summation of Behavioral Decision Theory:

> Human decisions affecting the future, whether personal or political or economic, cannot depend on strict mathematical expectation, since the basis for making such calculations does not exist; and that it is our innate urge to activity which makes the wheels go round, our rational selves choosing between the alternatives as best we are able, calculating where we can, but often falling back for our motive on whim or sentiment or chance. (Keynes 106–7)

Go and AI: Does AlphaGo make decisions?

Decision Theory is intended to make decision analyzable by mathematics where possible. It fully allows that valuation may be subjective, reason risky, heuristics misleading, and behavior irrational, but once the situation for a given decision is given, and we know statistically how unreasonable things are likely to get, the assumption is that it is still possible to systematize implications, steps and substeps, probabilities, and other parameters of decisions. If we just focus on what can be submitted to calculation in decision, we might think that the ultimate result of mathematical Decision Theory would be a program that made decisions by algorithm. And there is such a program that made a huge splash in 2016: Google's AlphaGo computer program for playing the game of go.

There have been computer go-playing programs for about thirty years, but until last year, none came close to playing at the professional level. Lee Sedol was asked to play a five-game match against AlphaGo because since the past decade or so, Lee has won more international tournaments than anyone else.

(To give an idea of the level of expertise involved: Beginners are ranked at about 25-kyu, working their way up to 1 kyu, then to amateur 1-dan [Master, similar to black-belt] rank, then up to amateur 9-dan level. [I am an amateur 4-dan.] If a player can become one of the strongest amateurs in the world by the age of about twelve, he or she may be invited to study with a professional in Japan, Korea, or China; after years of full-time go study, the very best of those may be promoted to professional status, beginning as a Professional 1-dan, and after five to ten more years, the very best of those will rise to Professional 9-dan, and the very best of those will win national and international tournaments.)

AlphaGo beat Lee Sedol 4 games to 1.

Does this mean that AlphaGo made decisions, or does it say something about what it means to decide? To think about the zone where AI meets phenomenology, we need to know a few details about how the program works. I base my discussion on the twenty-five hours of live game commentary by Michael Redmond, Professional 9-dan, broadcast on YouTube,[26] as well as on papers written by Google's AlphaGo programmers.

The go board is 19 × 19 lines. The game starts on an empty board. Players take turns placing stones on one of the 381 intersections. Once a stone is on the board, it does not move; if one or more stones are completely surrounded, they are captured, and removed from the board. The goal is to connect the stones into walls that surround empty space on the board. Once all the empty space is surrounded by the two players' groups, the game is over. The player who has surrounded the most territory wins. There are some additional tricky rules, and the strategies are profound, but the principles of the game are simple.

The AlphaGo program is designed with three modules. The first is the "policy network," sometimes called a neural net, which is concerned with pattern recognition. To begin with, games of go are entered into the computer's database. Programmers refer to this stage as "supervised training." To supplement the database of already played games, the computer is set to play games against itself ("training from scratch"), as many games at a time as the computer's processing can handle, twenty-four hours a day. The program then analyzes common moves and common patterns. With each pattern the game arrives at, the program searches its database of professional games and counts how many times a certain move has been played in that pattern. Then it assigns a probability to each move based on the likelihood that a human expert will play that move in a similar pattern. By considering only moves that have a relatively high probability of being played by experts, AlphaGo can eliminate most of the moves that can legally be played. On the 19 × 19 board, it would be impossible to consider every possibility for the 250 moves in an average game. The computer is fast, but not that fast; like its human opponent, the computer is on a time clock.

Prior to AlphaGo, the received opinion was that although brute force computers can calculate enough branching possibilities on the cramped 8 × 8 chess board to defeat the best humans, the 19 × 19 go board was too large even for a computer operating with fifteen Teraflops (one trillion floating point operations per second).[27] Computers evolved once programmers advanced from brute force to pattern recognition, like "Monte Carlo" systems that compare the statistical success rate of different strategies, eliminating sure losers using the mathematics of "surreal numbers" (which happens to be Badiou's favorite number theory[28]). Of course, humans too have to choose a move without looking at every possibility. In this respect, Michael Redmond says that AlphaGo plays like a human: Concentrating on only a few possibilities, evaluating the situation, and only then reading out just a small portion of the possible continuations.

At an early stage in Google's program design, the program was tested against amateur go players to see how well it would do if it used *only* pattern recognition and common pattern responses, and did not even try to analyze branching possibilities. It surprised both programmers and opponents by how well pattern recognition alone succeeded. Without reading out any branching sequences of future moves, the pattern recognition program defeated amateur players many levels stronger than any previous go programs, designed to look many moves ahead, had been able to do. The amazing thing was how well the pattern recognition program succeeded at go in spite of not looking into the future at all.

AlphaGo's second module is a "value network." Based on both the expert and the self-played games, the program surveys which moves in a given pattern result, in the final analysis, in won games. At this stage, the program still does not analyze, predict, or even look at the long sequences of moves that lead from the initial move choice to the eventual win—various games lead from that move to a win using different sequences to get there, and the computer program in this module does not consider these sequences—it only determines which individual moves played in a given pattern lead to eventual wins. It does not analyze the branching possibilities that for human players is what game playing is all about. This module improves AlphaGo's game results substantially.

We finally get to the search tree in AlphaGo's third module, the brute force algorithm that we usually associate with computer calculation. This module considers the opponent's follow-up possibilities to each of AlphaGos own possibilities, then AlphaGo's response possibilities to each of those, and so on. There are a few local tactical situations (josekis in the corners, tesujis for local best moves, life and death capturing problems, etc.) that are one-way streets once they get started, but most branching series have to be assessed move by move. For each move in the decision tree, AlphaGo repeats the procedures in the first two modules: Given a game position, which moves have the highest probability of resulting in a win (ignoring the moves in-between)? Each of those moves generates a new game position, for which again the computer evaluates which moves likely lead to a win. (Exactly how it balances the probability of a final win against the probability of local advantage is at this point a Google trade secret.) The program is indifferent to which moves are its own and which are the opponent's, so the procedure is the same for each move. In effect, the decision tree step is a multiplied iteration of the pattern recognition step. Computationally, this makes sense. What sense would it make to say that a certain next move is better than another, unless it is more likely to lead to a win? On what other grounds would the program select one path over another to analyze the follow-ups for?

Predictably, almost all the commentators and journalists talked about how AlphaGo "thinks," whether it was "worried" at certain points, and so on. I will limit my comments to particular features of the program.

- It is amazing that sophisticated pattern recognition alone (matching a single move to a given pattern) is sufficient to win games against strong (1-dan Professional) players, without any analysis at all of subsequent branching series. I cannot say too strongly how crazy this proven fact is, and how much of an effect it could have both for the nature of decisions and for the nature of time. We might interpret this in two ways: Either decisions do not need to have a futural reference in order to succeed in the future, as long as they recognize patterns; or else, patterns themselves are diagrams of the future.

- Pattern recognition can be used to define the best "next" move, but does not by itself define the next next, or the next after that. The program jumps from a present situation to the end situation, ignoring the process that would lead from the present to the end. It jumps from the next to the end, without the sequence in-between. It is the future that is omitted by this jump. In this sense, the program does not "look ahead." I suppose one might argue that human time-consciousness works like this too, that it is an illusion when we feel that by considering options and their consequences we are imagining our way progressively deeper into the future. But I do not see any reason to draw that conclusion. On my analysis of the future of decisions, a move without projection into a next move after that is barely a decision at all. It is not just that the next time without further continuation is limited to short-term futurity; it is not future-directed at all. The idea of a short-term future is interesting in its own right, but it is a different idea from the idea of the "next." We might allow that the "next" is decisionistic, a break from the present into something else, but decisionism as we have seen describes a situation almost the opposite of that of decision. If we define "play" as a future-directed noesis, AlphaGo wins, but does not play. (In Joe Haldeman's sci-fi novel *The Forever War*,[29] two planets hundreds of lights years apart are at war. Each sends a battleship with state of the art weapons to the other planet, but by the time it gets there, those weapons are hundreds of years out of date. The upside is that the defensive shields against those weapons will have rusted away hundreds of years earlier. The moral of the story is that acting toward the future by skipping steps is not impossible, but it does take a lot of the outcome out of the sphere of decision. With skipped steps, there is still a future, there is still a decision, and there is still a decision's future, but there is not much of a decision's future reference.) At any rate, no human would play go this way.

- On the other hand, what impressed strong go analysts is less that AlphaGo won so many games against the best human, or that it processes so much data, or that it repeats the best moves ever played by humans, or that it plays the percentages. What impressed the go world was that AlphaGo played "new moves," moves that had they been played by a human would have been called creative, brilliant, insightful ideas. The best move played by AlphaGo was in game 3, a shoulder hit on the fifth line. In everyday go strategy, a move on the fifth line early in the game gives the opponent room to play on the fourth line, and thereby to gain too much solid territory; playing on the fifth line makes a wall facing the center of the board, and this certainly has value, but it does not look like it will store enough potential to compensate for lost actual territory. It is true that there are professional human go strategies for playing toward the center, like Go Seigen's New Fuseki strategy in the 1920s, and Takemiya's "Cosmic Go" in the 1970s. So AlphaGo's strategy in game 3 did not look ridiculous, it just looked provocative, risky, interesting, and new.

- AlphaGo's unsuccessful moves, though rare, tell against its ability to "think." There were two occasions in the series when AlphaGo made mistakes that human professionals (even strong amateurs) would not make. In game 5, AlphaGo played a sequence that lost a few points on the side of the board. As Michael Redmond explains, it was a sequence that advanced go players learn, study, and play so often that they do not need to think about it. (It is interesting that a test of whether something thinks is tied to its capacity to think about something so often that it stops having to think about it.)

- AlphaGo played several moves into an obviously unworkable sequence, until it eventually broke off and moved elsewhere. Because the moves were on the edge of the board, there was no room for unexpected effects, so there is no way that the program had discovered something good about it that humans had missed. Of course, all players make mistakes, and even great players have what they call brain spikes. But it was a puzzle to the commentators to explain how a powerful analytic program with an enormous database of human games could have make a series of errors like this. As it happens, AlphaGo came back and won that game. But there was apparently some glitch in the program. A professional human would never bring herself to play such moves. Despite its success, AlphaGo would not have passed the Turing test on the basis of those moves.

- In game 4, Lee Sedol made a great move that won the game: A wedge tesuji that reduced part of AlphaGo's wall to a shortage of liberties, so that Lee's invasion stones could connect out of danger. Most professionals watching the game had not considered this move, and those that did could not see how it would succeed. AlphaGo did not prepare for this eventuality, and had to resign the game about thirty moves later. When the programmers looked at AlphaGo's code afterwards, they found that AlphaGo had determined that Lee Sedol's move had only a 1/10,000 probability of being played, so it did not spend much time reading that branch of the decision tree. But as both programmers and humans pointed out, a 1/10,000 probability move is not that small a probability in go terms, so AlphaGo should have spent more time evaluating it.

- AlphaGo's errors will be corrected by the programmers. Glitches will be easily rewritten for the next match, if there is one. AlphaGo's designers may have the program play a few hundred thousand more games with itself to look for new go strategies less and less patterned on human games. It is worth noting that while a human studying go full-time may play five games a day, a computer might play five hundred a day. Lee Sedol was 33 years old at the time of the match; AlphaGo was initialized two human years earlier, so had spent an equivalent of 200 human years playing go. When Lee is 36, and AlphaGo has been playing go for 500 human years, will that be a fair match? In any case, Google has not yet decided whether to challenge another human, or to declare the experiment over, freeze development, and sell the program as a toy to consumers.

- Journalists, as well as many amateur and professional go players, including Lee Sedol himself, treated AlphaGo's win as a blow to humanity. We are used to such blows to humanity at the hands of computers by now. Michael Redmond has a different view, which seems to me a good one. He wants to have AlphaGo on his personal computer to play with. The computer learned from humans, now humans will learn from computers, the computer will learn from its loss to a human and from its other mistakes, there will be many more discoveries about go both on the part of humans and on the part of computers, go strategy will improve, and humans will have lots of interesting new ideas to develop. Computers may not think, play, or make decisions in the sense that humans do, but the interaction between humans and computers will lead to a lot of human decisions.[30]

Li Zhe, a Professional 6-dan go player and the National Champion of China in 2004 and 2006, who left the Go world in 2012 to study philosophy at the University of Beijing, wrote a very interesting blog about the match.[31] In several ways, he says, AlphaGo works in a humanlike way, particularly by strategizing globally. However, it is not human to assess a move's "winrate," rather than by how many points it will get. Analyzing a sequence is "logic," whereas analyzing win rates is statistics.

Technically, we know more about how computers "decide" than how we humans decide. We do not know what program, if any, is running in us. The problem is that knowing technically how computers "decide" does not mean that we know whether they decide. Li Zhe asks good questions: "For AI, what is momentum? What is bravery? What is the killer instinct?"

My purpose in raising the case of AlphaGo is not to conclude that future-directed decision either is or is not computational. My purpose is to expand the range of structures that might describe different sorts of decisional futures. AlphaGo got strong by dispensing with the temporal middle ground between the present and the distant future. The interesting thing is that there is no absolute division between present and future in its strategy. For the program, and possibly for us too, between any move and the end of the game, there is always a potentially ignorable zone, a potential discontinuity in the series, which nevertheless could by decision be refilled with as dense a series of alternative futures between any one point and any other as there is time to analyze. Without a series, there is, I still think, no future. But the structural variations on seriality can afford to be expanded when we see a nonhuman logic of seriality get somewhere good. Humans have our own ways of referencing futures without mentioning all the steps along the way, of planning for futures without solving all problems in advance, of imagining futures without having any idea how to get there, of displaying futures in leapfrogging montage sequences, and of plotting future time lines with unevenly dense subspaces (subtimes) for alternatives. AlphaGo will push human decision-makers to add variations to their own temporal habits.

The Google DeepMind programmers treat go as "a game of perfect information" (Silver 484),[32] by which they mean primarily that there is an unambiguous winner and loser. But success in programming is still measured at each stage by fuzzy sets, stacked probabilities, and time limits on sequential sub-decisions. DeepMind takes pride in its novel approach to backward induction, or "backpropagation,"[33] which minimizes "lookahead search" (484). It does make use of branching, but prides itself on having evaluated thousands of times fewer branching positions than IBM's Deep Blue evaluated when it beat Gary Kasparov in chess (489). This goes to show that branching is just one subroutine involved in getting a decision from the present to the future and back. Even so, we need to know what it means to branch.

CHAPTER SEVEN

Branching futures: Tense logic and multiple worlds

It sounds natural to say that decisions cause events to branch into different possibilities. There are various ways to say this: Decisions are possible because events branch into different possibilities; decisions select among branching futures on the time line; decisions make events turn in a new direction; decisions follow one path rather than another; decisions leave some versions of our life untraveled; in decisions, we branch out.

The image of branches on a tree suggests a more determinate differentiation of branch contents and discrete pathways than I have in mind. To articulate the multiple future realism I am aiming at, it would probably be better to talk of rhizoming than branching futures. We should always be skeptical about a metaphor, particularly one that uses a spatial figure to describe a temporal structure. One might even agree that decision-making assumes that there is a branching future, but not agree that the future actually branches. But let us not draw that conclusion unless we are forced to. What is meant by a branching future?

To begin, we should distinguish branches from "forks." A "branch" divides a single-event series into several independent lines after a turning point. A "fork" is the converse: A point of interaction at which previously independent-event series get entangled into a single series (Price 138–40).[1]

In some ways, branching futures are similar to possible worlds, but in other ways, not. Possible worlds and branching futures both conceive of a plurality of possibilities, and both pluralize the virtual world. The difference is that possible worlds are generally treated as different from each other from start to finish. For Leibniz,[2] God makes his choice among possible worlds at the moment He creates the world; once this world is the actual world, it unfolds as only this world does—there are no branching futures within a given world. One consequence of the independence of possible worlds is emphasized by David Lewis[3]: No actual entity can belong to more than one possible world. A person may have almost identical "counterparts" in different possible worlds, but no individual can exist in two different possible worlds, or interact with a world other than her own, or force her actual world to branch into a second world (see Chapter 9.) Possible worlds are not distinguished by a splitting off in the time line the way branching worlds are. In short, pluralizing the world modally and pluralizing it temporally are different, just as it is different to think that there is a plurality of worlds, and that each world becomes plural by itself.

The difference between possible worlds and branching within a world is often blurred in science fiction scenarios that combine alternate worlds and time travel. In the film *Primer* (Shane Carruth 2004), the character goes into the past and keeps living there until he reaches the present he left from, which means that in the present, two versions of himself exist. In films like *Looper* (Rian Johnson 2012, or the *Terminator* franchise), a character comes from the future to change the present so that the future will be different than it was/will be; he may return many times in hopes of getting the change right. In all these cases, time travel leads to alternate worlds, but this combination is not quite the same as branching, where a single choice splits off into many worlds. In time travel paradoxes, a single world ends up including contradictory states. In modal realism, there exist many possible worlds. In branch realism, an event *generates* many actual worlds.

Branching assumes temporal asymmetry, so we should say a word about this. In the philosophy of science, as well as a lot of analytic philosophy of time, there is a common view that time is symmetrical, that there is no objective difference between the direction of time leading to the future and the direction leading to the past (or between a branch and a fork). Some arguments for time symmetry are based specifically on special relativity.[4] But well before relativity, the argument for time symmetry was based on the fact that a scientific law is about the evolution of a system, and tells us with equal degrees of certainty what the states of that system were in the past and what its states will be in the future. As it predicts both directions equally, the argument concludes that the past is not more determinate, or more knowable, than the future. And as the most common way of distinguishing past and future is that we know the former but not the latter, the very distinction between past and future is undermined. Furthermore, if a system can evolve in either direction, there is no particular reason why we should expect our world to evolve in one direction over the other, that is, why it should, or even that it does, evolve toward the future and not to the past. Some versions of time symmetry concede that although there is no objective difference between past and future, subjectively we experience what feels like a difference; others do not concede even this. Some versions find exceptions to time symmetry in exceptional natural phenomena, like entropy, or particle decay due to weak nuclear forces, or collapsing wave functions. Physicists generally take such natural exceptions to show that time symmetry is broken, but philosophers like Paul Horwich[5] and Huw Price try to get around those exceptions. Other versions find special exceptions to time symmetry in moral phenomena, like agency and decision-making. Some philosophers of science reject time symmetry altogether, arguing that although laws make symmetrical predictions, causal sequences in the physical world are asymmetrical.

For the purpose of this book, sadly, I am going to block out the question of cosmological time symmetry. In this chapter, I touch on resources for the theory of branching futures: Tense logic, Many Worlds interpretations of quantum physics, and time travel science fiction.

Tense logic

Aristotle

Aristotle's *On Interpretation* section 9[6] considers whether statements about the future are all either necessarily true or necessarily false. If a statement is contingently true (or contingently false) at a certain time, it seems that it would be necessarily true (or necessarily false) at all times after that. For example, if it is contingently true that "It is sunny today/June 23, 2015 in Pittsburgh" it would for all time afterwards necessarily be true that "It *was* sunny on June 23, 2015 in Pittsburgh." The future will keep saying things we already know are true. And similarly, for all previous times, it would have been true that "It *will be* sunny on June 23, 2015 in Pittsburgh." Whatever is already true in the future, as it were, is something we could already say now. In both cases, we compare present and future propositions that say exactly the same thing with different tenses. This point seems plausible enough as long our only concern with truth in the future is that the future can formulate what happened before it, and vice versa. But, if we do not know which propositions will be enunciated in the future, and if we do not know whether propositions enunciated now about the future will correspond to anything in the future at all, how do we analyze their truth or falsity? What kind of truth or falsity, if any, attaches to sentences that we utter about the future now, before the future events are anything? Is there some kind of being in the present that continues into the future, so that we can use it to verify sentences about the future? Aristotle's view is that generally there is not sufficient continuity of being for that. Therefore, sentences about the future are generally contingent.

Aristotle's explanation is quite good. "In those things which are not continuously actual, there is a potentiality in either direction" (19a). In the rare cases where an event is continuously actual, or in other words, when its potential leads in only one direction, then its momentum does not include any branches. For continuous, unbranching topics, the truth or falsity of sentences is independent of tense. If a sentence of that type in, and/or about, the future, is true, then it is necessarily true (and if false, is necessarily false).

In contrast, when an event is not continuously actual, that is, when it is actual but discontinuously so, then it has potential in many directions. What creates plural potentiality is discontinuity. We can apply this nicely to the way a decision works in more than one direction: A decision's effects are actual, but not continuously so. A decision's follow-through is discontinuously actual. And because of this effective discontinuity operating through future time, it operates "in either direction." Aristotle does not envisage that an event or a decision can actually operate in many directions, but he does envisage that its future potential branches in many directions.

A coat might or might not wear out, a sea battle might or might not take place, and in each case both directions are potentiated by the same present conditions. The present state and the future direction it takes are discontinuous.

Aristotle anticipates a central problem in modern tense logic regarding disjunctions in the future. We might be tempted to say that disjunctive propositions about the future, propositions like "Either there will be a sea battle tomorrow or there will not be," are neither true nor false, since at the time the proposition is stated, nothing has happened one way or the other. Neither of the disjuncts corresponds to anything, so the disjuncts are neither true nor false, and for a disjunctive proposition to be true, at least one of the disjuncts must be true. Therefore, disjunctions about the future are not true. (Indeed, on this argument, they should all be false.) Furthermore, if propositions about the future are neither true nor false, then Disjunctive Syllogism (DS) will not work on them. Compare this situation with what we would normally expect for a proposition of the "P v ~P" form when P is asserting something about an event at future time-2. Normally, we would expect that it might come about at time-2 that the event P does not occur, which makes it the case now at time-1 that "It is not the case that P at time-2 will be true." If DS works, then it should follow that now at time-1, "It is the case that ~P at time-2 will be true." In other words, we expect "P v ~P" to hold for future propositions, so that if P is false, then ~P is true. It seems hard to accept that DS would not work for propositions about the future, because if DS breaks down, quite a lot of logic might fall with it.

There are two difficult, but equally interesting, alternatives. One is just to accept that disjunction does not apply to the future. The other is to accept that the principle of noncontradiction does not apply to the future. The latter is not out of the question: It is conceivable that two contradictory potentials are equally affirmable so long as the event is future. Aristotle clearly rejects the former, that is, that disjunction does not apply to the future: "It is necessary that it [the sea battle] either should or should not take place tomorrow" (19a). Instead, he appears to carefully accept something like the latter, although without rejecting the principle of noncontradiction altogether: "When in future events there is a real alternative, and a potentiality in contrary directions, the corresponding affirmation and denial have the same character" (19a). I do not know how to judge whether one of these solutions is ultimately better than the other. I think that removing disjunction from the future captures the fact that the future is not divided up in the same way as actual events in the present are. But I also think that inclusive incompatibilities capture the fact that the future covers different projections over the same temporal field. I do not think we need to decide here which peculiar concession we have to make in tense logic. Different systems for formalizing logical relations can sometimes express the same realities. What is important is that the future has peculiar characteristics that cannot be captured by the same logic that captures present and past propositions. The future requires a kind of truth-function where some propositions that do not correspond to facts nevertheless count as true; or where some propositions that do correspond to facts may be consistent with others that contradict it.

Agency

The nature of future sentences, and sentences about the future, changes structure when the sentences are performatives, as when the future is not just judged, but decided upon. One might have expected that agency would be a rich resource for theories of decisional futures. But there are natural reasons why a theory of agency might regard decision as a relatively small subtopic. Agency is in some ways broader than decision, as it involves stages prior to decision (character, habit, etc.), as well as posterior to decision (causal interaction, indirect consequences, etc.). Yet decision is in some ways broader than agency: The future of a decision branches into unintended futures as well as those covered by agency. Still, with the right *mutatis mutandis*, some theory of agency might lead back to decisional future branching.

Belknap and Perloff[7] propose an account of agency that emphasizes (but not too much) the agent's causal power, while also emphasizing (but not too much) the agent's intention, and also (but not too much) the futural sequence of events initiated by the agent. Rather than say an agent "causes it to be the case that ...", they prefer the weaker formula, that an agent "sees to it that ...": "STIT." I cannot wholeheartedly vouch for this formula, but it has the virtue of ambiguating in a realistic way the relation between a decision and its future. When we say "*a stit: q,*" it does not matter whether the agent *a* is singular or collective, or whether the event *q* is momentary or historically spread out over sequences or branches. Because it allows action to see to futures without intentionally causing them all, STIT theory's "picture of the future as replete with possibilities is expressed in a theory of branching time" (Belknap and Perloff 189).

Thomason on truth-value gaps

The theory of tense logic for future statements that Belknap and Perloff favor is Richmond H. Thomason's influential "Indeterminist time and truth-value gaps."[8] Thomason attempts a semantic for alternative futures that does not reduce them to epistemic unknowns, but treats branches as part of real, multi-linear, time (Thomason 265).

Thomason's model is that each moment *m* can be followed by two (the exact number should not matter) possible moments. More precisely, the moment following *m* can be "filled" in two ways. Because one moment does not determine which of two possible next moments will occur, the branching future is indeterministic. Histories "pass through" moments. An agent at *m* has a "choice set" for how to proceed to the next moment, and how to follow different long-run continuations. This does not assume that there are payoffs for good decisions, so it is not necessarily tied to Decision Theory; all that choice-time guarantees is that "the world goes on" (191).

For Thomason, until the future moment *m* occurs, a proposition P about what happens at *m* has no truth-value. If we look at the set of all true propositions, and the set of all false propositions, we will not find the proposition P in either set. The union of the two sets ought to include every proposition, but there are gaps in the union set where propositions about the future should be. Does this entail, as Aristotle worried, that the Law of the Excluded Middle does not apply to propositions about the future? The theory of truth-value gaps for branching futures is supposed to manage problems like this, by analyzing truth-conditions for future tense.

It seems to me that Thomason's model requires questionable assumptions: that there exists a set of times uniquely ordered by a before–after relation; and that every event occurs at a moment

(There is no reason why tense logic could not locate events during intervals, 265). For Thomason, it is obvious that a time must have a unique past; different past times would necessarily be followed by different successor times. But must that be so? Could two worlds not have different past histories that converge at one point, which is thereafter the same time for both worlds? Does something like that happen when an undecided person resolves on a turning point?

These questions aside, it is interesting to consider a logic of future tense for branching alternatives. Should we say that a proposition P about an event in the future at time-2 has no truth-value now at time-1, but that it will have a truth-value at time-2? If it is possible now at time-1 that P might be true at time-2 and also might be false at time-2, should we say that there are two branches from time-1, and that on one branch P is true and on the other branch P is false? Should we say that if P is a tautology, so that it will be true in every branch of the future, then it is necessarily true not only at time-2 but already at time-1? That would mean that some propositions about the future do have truth-values, indeed that the only truths about the future are necessary truths.

The key to solving these questions is whether, when we consider truth-value gaps, we imagine that when we get to the future, when something happens, we will be able to fill those gaps with truth-values, or not. If truth-value gaps are not permanent, but can be filled in later on, then they are never outside the scope of truth-value altogether. But another way to think of this is that truth-value gaps may be permanent. It may be that the set of future-referring statements we make now will always contain truth-value gaps, even if we will make other statements in the future that will have straightforward truth-values at the time they are made. In other words, what happens to truth-value gaps, and to future branches of events, when the future arrives?

Broad and McCall on branches

There are many ways of imagining the evolution of branching futures over time. A great paper by Rachael Briggs and Graeme Forbes[9] compares C. D. Broad's "Growing Block" model of the future with Storrs McCall's "Shrinking Tree" model. I follow their lead, and make my own comparisons between Broad and McCall.

C. D. Broad, in the 1920s[10] (with connections to McTaggart and Bradley), took up the old problem of whether past, present, and future have the same or different ontological status. Does the same event endure, but change its ontological status when time passes over it? As Broad says, we clearly do not think that different slices of an event exist in different parts of time; rather, we think that "the whole event was future, became present, and is now past" (Broad 64). But of course, past, present, and future are not three different worlds, each of which has the whole event in it. Broad's view is that when a present event becomes past, the event's characteristics do not change; the event just gains relations to the events in the next present, that is, to the events that came after it, that is, to the events that it preceded. When a new event occurs, the previously existing events now have a new event to stand in relation to. Time, Broad concludes, is not a measure of anything on its own terms, and is not a kind of flow or medium in which events occur, but is just an asymmetrical way of describing an event's relations to other events. Assigning a time to an event simply clarifies which other events are related to it. Different times are just different sets of related events, and all periods of time, despite their asymmetrical linearity, have the same ontological status. "The past is thus as real as the present" (64).

As the future consists of adding new events for existing events to be related to, it presents a bigger problem for ontological status than the past does. Past and present describe the relation of events to events they succeed (66). How are events related to events that succeed them, that is, to the future? The whole idea of the future is that it covers events that do not exist yet; by definition, the future is at issue only as long as it does not exist. For Broad, "future events are non-entities, they cannot stand in any relations to anything, and therefore cannot stand in the relation of succession to present events" (68). Future events do not in any sense exist, and if they have a temporal sense at all, it has nothing to do with having a successor relation to other events. Broad concludes that the future thus has nothing to do with "change," causality, or even relation, although he concedes that there might be some concept of mere "becoming" that would suit the description of a nonexistent future (66–7).

This calls for tense logic. Sentences about the future seem to be about something, but that is not the same as referring to a fact: There is no fact for a judgment about the future to refer to (72). Therefore, at the time they are made, judgments about the future are "neither true nor false. They will become true or false when there is a fact for them to refer to; and after this they will remain true or false, as the case may be, for ever and ever" (73). Broad concedes that the Law of the Excluded Middle does not hold for judgments about the future. If we wanted to give Broad's law for future propositions a name, we might call it a Law of Nothing-But-Middle, or a Law of All-Inclusive Exclusion.

All of this constitutes what Briggs and Forbes call the "growing block" theory of time: "The sum total of the existent is continually augmented by becoming" (69). It might not be branching, but since events on this theory are fed into the time line from the future end, it is a kind of time funneling (like Bergson's memory cone[11]). One new event at a time enters the picture, through a small hole at the entrance, and time present and past fatten up with new events and their relations.

The contrasting picture is Storrs McCall's "Branch Attrition," perfected in the 1990s.[12] On his picture, the past consists of determinate events that occurred one at a time, so the past is a single linear straight time line. The present is a determinate point from which future branches extend. Each future date consists of a plurality of possibilities, which branch away from the present in different directions (McCall 1–19). As time passes, the point of the present moves up the branching tree. All but one of the many future possibilities for a certain future date disappear once that date finally becomes present and an event occurs on it. That is, as soon as that date, once future, becomes present, and then becomes past, all but one of the branches on which that future date used to be found, vanish, and that date is now a single point on the one and only branch that remains, which henceforth is the linear time line of the past. This is "branch attrition": As events unfold, all branches but the one actual event on the time line of the past are eliminated.

Briggs and Forbes prefer the growing block to branch attrition. One reason is that they picture that when future branches disappear, some otherwise future children will "drop out of existence" and be destroyed, which for Briggs and Forbes is a moral reason for rejecting the ontology. Their other reason is that space–time futures are more of a burden to justify than "an ordered set of possible worlds" (Briggs and Forbes 8); they think it is more of a burden to justify the existence of branches of the future than it is to assign meanings to propositions about the future (32). In their words, it is easier to justify "ersatz futures" than real futures. And a theory that is easier to justify is, by razor, more justified. (For my part, I am skeptical about this calculation of burdens.)

There is a chapter at the end of McCall's book on the topic of decision, but it focuses on how indeterministic choices might be compatible with responsibility, rather than on how decision contents are defined in branching terms (McCall 257). How might we apply branching futures to the problem of decisional futures?

To begin with, do we have to believe that future branches undergo attrition once a single actual event takes place? Why can the non-actualized branches not still branch on a virtual time line—the same virtual time line they were already on while they were still future? No doubt, in an obvious sense, the actualized branch becomes isolated on the actual time line, but could the non-actualized branches continue to exist with their own status, a virtual status that parallels the actual time line, a status that could one day become actual in a different present? Can the actualized branch and the non-actualized branches all have the same status qua time branches? If that is the case, then since the present is not a point at which branches undergo attrition, can the past branch out in the same way that the future does? And applying this to decisional time, can a decision content continue to branch into options even after the decision-maker has fixed on one particular option to pursue? This is the model I have been suggesting: reversible, hesitant, spin-off decision-making that conserves unchosen paths. Can a model of branching futures be adapted to this structure?

As one might expect, there are philosophers who think that the whole idea of future branches is unnecessary. According to Paul Horwich (30), we only think that the future branches because we do not now know how to verify or falsify propositions about the future. Branches represent a lack of our knowledge rather than a structure of time. In the future itself, Horwich figures, there will be evidence for and against propositions regarding what is now the future. If we take the long verificationist view, it will always turn out that the future is not branching at all, since only one so-called branch will ever be verified.

There are two ways to defend branching in the face of the argument that branches only represent epistemological limitations.

One way is to say that statements about the future are not primarily statements about events when they will occur, but are essentially statements about events before they occur. So even if events are verifiable as soon as they occur, that is, even if events are verifiable as soon as there are no alternative branches, it may still be the case that statements about the future refer to future branches until such time as those events do occur. Branching may still be part of any time series where we talk of events in the future and not just events in the present. That is the point of future branches anyways.

The second way to preserve branching is to say that the meaning of statements about the future is not primarily about how we verify them, that is, how we observe future events, but about how we make decisions regarding what those events ought to be. In that case, branching is part of any time series described as a decisional noema, that is, part of any time series synthesized by desire and not just perception.

Dummett on past and future: Realism versus anti-realism

The problem of whether future branches belong to semantics or to ontology is subtle. Michael Dummett analyzes realism and anti-realism in relation to past time, but we can apply his questions to future time.[13]

The trick is this. When do we assess the truth conditions of a statement about the future: While the future event is still in the future (while the proposition about the future is still neither verifiably true nor false), or after it has occurred (once the proposition is settled as true or false)? On the one hand, it should be while the future is still future, since we mean to talk about the future; on the other hand, it should be once the future happens, about what actually will happen, in the future. Which way treats the future as future? (a) Say now: P will be true at a future time, or (b) Say at a future time: P is true now?

Dummett says that (a) *seems* anti-realist with regard to the future, since it interprets the future in terms of what we say and know in the present; (b) *seems* realist with regard to the future, since it posits that some facts in the actual future will make P true. But Dummett argues that there is no real difference between these two options for a person who exists in time. That is, if we think of ourselves *in* time *now*, so that everything we say is said now, then saying now that P will be true in the future is just the same as saying that if someone in the future says P, it will be true then. On the other hand, *if* we were *not* in one time, if we were "outside the whole temporal process" (Dummett 369), if we could say the same statement from the perspective of several times at once, and compare their meanings, then there would be a difference between our saying something in the present about the future, and saying it in the future about what will then be present. This generates a twist that might surprise a Heideggerian. Being neutral about which time one is in, being *not-in-time*, being anti-realist about being-in-time, is what allows us to make a realist distinction between present and future. In this sense, "it is the anti-realist who takes time seriously" (370).

This is a subtle point, as one would have thought that if there is a real distinction between present and future, then only the people in the present would be able to experience the present as present, and only the people in the future would be able to experience the future as present. But problems with tense generally arise when the time of an experience is too rigidly fixed at a temporal location. The experiencer who is anti-realist about her own temporal moment does not experience just that one present moment; neither does she experience many temporal moments within the framework of a single moment, as a *totum simul*. She carries the past and moves into the future all the while. This, after all, is what *durée* is about: To experience the future not as some imaginary other person's present, not as the future anterior, not as what will have become present; but as the future of the time one is living through now.[14]

In short, the ability to see the future differently from the present requires that we see and compare different times at the same time, that we think an event from different temporal perspectives at once. Dummett articulates this as being outside time, but it is better to formulate it as being inside multiple time positions. The latter sounds more like the Deleuzian picture, in which the coexistence of times overlays the succession of time. As we will see in Chapter 9, Deleuze takes time seriously precisely by not taking the usual succession-oriented account of time seriously. When Deleuze and Guattari say that Nietzsche is right to identify with all the names of history,[15] they are recommending that a speaker, or a collective assemblage of enunciation, speak from many different times at once. This makes Deleuze a realist with regard to the future, in the sense at hand, in that different future positions are really different for a subject who speaks from all of them.

Obviously, if a person conceives an event from irreducibly different temporal perspectives at once, it will be more difficult to specify which objects and which time references are in a person's consciousness at the time they make a decision. Indeed, it will be more difficult to identify the "person" seeing that event. But these are the right complications to face. They require personal identity to be laid out in what David Braddon-Mitchell and Caroline West[16] call "temporal phase pluralism" (which sounds a bit like the temporal event-slices that Broad rejected earlier). Assume that a person can make a decision that gives her a different personality, and so turns her into a different person, or a different sort of person. Has the original person survived or not? For Braddon-Mitchell and West, a person exists as a series of phases, and by decision, enters and exits some of them at different times.

One consequence is that a person-stage could conceivably be a part of more than one person, that is, a person can be two persons over time, and during an interim period of time, a person can be both persons. Another consequence is that when a person tries to decide whether to make a life-changing act, and become a different person, there is no rational way to assess what she will want to have happened once she will have become the other person (Braddon-Mitchell and West 60).

In an odd way, Norbert Wiener's cybernetics[17] likewise requires temporal periods during which actual situations and data sets are ambiguous. If we had to think of a sequential decision as taking place over several slices of space–time, each with its own data to go by and its own judgment, the sequence would look like a series of closed world-sets, each one branching into autonomous futures. But if we think of sequential decision as a feedback system, where there is an ongoing interchange of messages between the decision-maker and the world (such interchange is called "information," 17), then the decision sequence looks like the lived duration of a perceiving organism. Wiener refers to the feedback procedures in a living organism as its "central decision organs" (30). It is interesting that cybernetics provides an argument against data sets in favor of virtual life. The point is that the metaphor of branching does not have to presuppose snapshot-like decision instants; once instants give way to durations, branching refers to something more like phase-pluralism. "Branching" may no longer be a very apt metaphor for this, but all metaphors eventually lose their appeal.

Flashforwards and time travel

The problem of referring to either past or future is unavoidable in any form of snapshot representation, since presumably all representations, whichever times they designate, are experienced in the present. If decisions were representations of the future, they would share this problem with paintings, stories, and films: How would they specify that they refer to future events, given that they are experienced only in the present? How is either the past or the future narrated by a narration in the present? A literary representation has the resource of tense, and can simply "say" that it refers to past or future, even if it is difficult to explain exactly how that works. In contrast, paintings that are trying to represent historical subjects have few resources for saying that they represent the past and not the present, other than to depict images that we know look like Napoleon or some such. Films have something in-between the strong resources of language and the weak resources of vision: They have flashbacks and flashforwards, or what Maureen Turim calls "temporally disjunct inserts."[18] How do flashbacks or flashforwards communicate that we are seeing something in a different time order than the time order in which we are viewing the scenes when we watch the movie?

A common feeling is that when we see a moving image on screen, the event depicted is happening within the narrative of the film in present tense. But Gregory Currie's *Image and Mind: Film, Philosophy, and Cognitive Science*,[19] argues (in part responding to David Lewis' possible worlds theory) that we cannot simply assume this "Claim of Presentness." Currie runs through several possible theories to explain how we can see events as happening in a different order than, or anachronous in relation to, the order we see them in. (a) An image may be a "sign" that gives us information about when it happens/happened (Currie 202). (b) The *viewer* may "imagine herself to be a time traveler," seeing things that happened earlier (202). (c) The viewer may suppose that the *character* onscreen in the present is remembering a past (203). (d) The viewer may suppose that an imaginary *narrator* in the present is telling a story about the past (204). (e) The viewer who sees a character performing one action in one scene, and a different action in another, may be forced to construct a temporal order for them (205). Currie finds all these solutions implausible, depending as they do on the presumption that viewers perform cognitive acts that we do not generally experience when watching films.

Currie's preferred solution is that the "proper treatment of anachrony" is instead based on McTaggart's B-series.[20] McTaggart famously argues that time is more likely ordered by objective before–after relations than by what he argues (I will not replay those arguments here) are spurious and contradictory subjective past–present–future relations. Ultimately, McTaggart thinks that before–after relations are spurious too, but many philosophers influenced by McTaggart ignore that. In any case, Currie's version is that events have before–after relations independent of how a subject might experience them in the present (206). We cognize that one scene represents an event dated at one time and that another represents an event at a time dated before or after it. The before–after relation will be in the meaning of the events, and not in a special kind of experience of the viewer. The advantage of this solution is that it does not require any special activity of the viewer that distinguishes how we view flashback scenes as opposed to how we view non-flashback scenes. We do not need to time travel into our own pasts or futures, or to suppose that either a character or an imaginary narrator is having an out-of-present experience. The inadequacy of this solution, though, is that it does not really explain what sort of temporal objects the "before" and "after" are, or how a before or after can be cognized when a flashback or flashforward is onscreen.

What is interesting to me about cinematic time-flashing, and difficult to explain, is the way a transition from a determinate scene set in the present, to another determinate scene set in past or future, is generally carried out by a simple cut. There is no shot in-between two shots that says that a temporal transition has been made, there is only an instantaneous and invisible splice. The futural reference of a decision shares this with flashforwards: it is carried out by splices and hinges, and otherwise has no visible mark. Decisions are branch points, branches are discontinuity points, and aside from their referential logic, discontinuities are nothing but cuts sequenced by montage.

This idea of temporal montage is at the heart of Pasolini's analysis of breaks in the time line. For Pasolini,[21] both film and life depend on time being broken into segments with different values. In principle, we could imagine two kinds of open future: A future without focus points, and a future with branching foci. If the open future is an unbroken continuity without focus points, it will be decision-free. There will be nothing morally decisive, or personally meaningful, in it. This sort of continuity may be "natural," but a value-making consciousness cannot live without breaking sequences up and valuing one branch over another. Pasolini makes the Heideggerian point about finitude: "Hence the reason for death. If we were immortal, we would be immoral, because our example would never have an end; therefore, it would be undecipherable, eternally suspended and ambiguous" ("Living Signs and Dead Poets," 248). It is the cut-off of time segments, and not their openness, that gives decisions their urgency. Cutting off a time segment does not mean it is closed, but rather that it is unfinished.

For Pasolini, what distinguishes different ways of being in time, and so distinguishes one school of filmmaking from another, boils down to something as simple as the "length of shots" it separates time into, and the temporal "rhythm" among the lengths of those shots ("Is Being Natural?" 241). For example, the ultra-long continuous shot in Warhol's film *Sleep* (1963) could be found in no other cinema—not because of its style, but because of its metaphysics. Pasolini says that he could never shoot in that "insanely naturalistic way." Even if the subject of a film, or an experience, is insignificant, Pasolini wants to show how "that which is insignificant, is," whereas Warhol's uncut temporality claims that "that which is, is insignificant" (241). In other words, for Pasolini, time-consciousness is underutilized when we watch a sleeping man's body undergo involuntary twitches for a long time. Continuity is merely "the accidental passage of time—the unreal time in which what is organic wastes away and runs down" ("Quips on the Cinema," 227). Decisive time-consciousness, in contrast, breaks a sequence down into segments, edits gaps in the time line and in points of view, inserts micro-deaths, and therefore moments of value, into the longer sequence. In short, the future is segmented rather than smooth.

The structure of segmented futures thus belongs to speculative imagination as much as to tense logic. The literary genre that deals most with the future is of course science fiction, particularly the subgenre of stories about time travel to the future. The American 1950s radio show "Dimension X" began each episode with the tag line: "Adventures in time and space, told in the Future Tense!" Disappointingly, they are not actually told in future tense, but are acted out in present tense. Almost all science fiction narratives are written in past tense, as usual. A few are written in future tense, like John Chu's fine story, "Thirty Seconds from Now."[22] In this story, the future tense narrative is motivated by the protagonist's ability to experience branching future possibilities before they happen. This is not ideal for the expression of the future as such, because while the story is narrated in future tense, it represents the present state of the protagonist's awareness. Indeed, this might be an implication of the future tense in general, that it is enunciated in, and indexed to, the present. The future as such might require something other than future tense to express it.

In fact, most science fiction forward time travel stories have the protagonist go into the future, and then present the future as if events are taking place there in the present (the future's present). This cannot be right, since the future is precisely what has not taken place. One of the few time travel stories to take this into account is Charles Simak's novel *Time is the Simplest Thing* (1961)[23]: When the protagonist goes into the future, he finds it empty. Nothing is happening there at all. Obviously. (When he goes into the past, everyone there is already dead. This is less obvious; the past should be empty too, by now.) The same problem was noticed earlier in an anonymous letter to the editor of the pulp science fiction magazine *Amazing Stories* in 1927, challenging H. G. Wells' *Time Machine*: "'How could one travel to the future in a machine when the beings of the future have not yet materialized?'"[24]

There are a lot of stories, like John Brunner's *Timescoop* (1969),[25] where a time machine brings an object from the past into the present. Although these stories are about time travel to the past, they pose paradoxes equally for time travel to the future, since the present is the future for that past object. Insofar as the present, its future, did not exist for that past, there should be no future for the past object to come to. Nothing from the past (not even the perceptions we think we remember) should be able to get to the present, which does not exist for it. Paradoxically, though, that object's future is our present, so we know as a matter of fact that the past's future does exist.

We cannot rely on the old trope that the future, when it happens, will be the present. Of course, it is true that one day our children will grow up, but is it right to say that what is in their future now will one day be present reality? If it is correct that in some sense the future is empty, then whatever happens to our children, it will not be the future. No doubt, the image we have now of our children as adults will one day bear either a close or distant relation to what will at that time be their present. The point is that the future is not a present that has not yet taken place, any more than the past is a present that is no longer taking place. In Chapter 9, we will consider Deleuze's controversial analysis of the future as the empty form of time.

Paul Nahin's *Time Machines: Time Travel in Physics, Metaphysics, and Science Fiction*[26] contains an exhaustive categorization of time travel science fictions (up to 1998), as well as a series of Tech Notes describing time travel technologies imagined in physics, from time dilation machines, to Tipler's infinitely long rotating cylinder time machine, Gödel's rotating universe with tilting light cones time machine, black hole time machines, wormhole time machines, and cosmic string time machines.

Nahin complains that J. J. C. Smart[27] and other philosophers (all Nahin's philosophical sources are analytic), instead of consulting physics, offer a lame semantic argument against time travel (Nahin 256–8). The lame argument is that if a person on April 12, 2017 instantly travels 100 years into the future, the date on arrival is both 2017 and 2117. Smart thinks that is a contradiction; it is not. Or again: If a person from 2017 travels back to 1817 and changes the past, when did the past change: in 1817 or in 2017? Or again: if 1817 is now changed, so that the original 1817 events *never* took place, how is there anything that the changed 1817 is a change *from*? Nahin thinks that if we restrict time travel to *observing* the past, but do not allow that it could *change* the past, the contradictions go away. Indeed, there is no contradiction even if time travel "affects" the past, that is, determines for the first and only time what happened in the past. (A modern character time travels back to the ancient world and poisons Alexander the Great—that is the only history of Alexander there ever was, it just happens to have been caused by a time traveler). But there is a contradiction for time travel to "change" the past, that is, to determine in a second way something that had already happened, which would make the past be two different things at the same time. (A modern character knows from history that Alexander lived to old age, travels back, and poisons Alexander instead, thereby causing a second history of Alexander).

The situation for the future is still more of a puzzle. If two people at different times in the present travel to the same date in the future, and perform incompatible actions there (whatever "there" means), then what will the future actually be once it arrives? Of course, the problem can be nipped in the bud if there is no such thing as time travel to the future, or if the only way to "travel" to the year 2030 is just to wait until it is 2030. But the other way to solve the puzzle would be to allow that the future can be more than one thing, that there can be different time lines projecting into and composing a multi-rayed future.

Nahin, who is trying to allow for as much time travel as he can, concedes too much, by conceding that the past, and hence the future, cannot be more than one event series. To put it bluntly, why is it a contradiction for there to be two time series?

The Many Worlds interpretation of quantum physics

There is a way to think about multiple pasts and futures in terms of the admittedly controversial Many Worlds interpretation of quantum physics.

I should say that I have at best an amateur layman's understanding of physics. Philosophers should be extremely careful not to claim more knowledge than they have. It is natural and desirable for us to want to draw resources from any field we find them in, and to speculate on how ideas might be adapted from one field to another. But if readers of these pages should find the conceptions of virtual time lines in physics interesting, they should seek out the sources for themselves, and make their own judgments.

The Many Worlds interpretation of quantum physics is an alternative to the collapsing wave-function interpretation of the Copenhagen school. According to the latter, certain properties of particles are indeterminate, but their probabilities are measured by a wave function; when the particle is "observed" (any event in which the particle plays a cause–effect role can count as the observation), the probability wave-function collapses, and the particle is then measured (the measurement is the observation), and found in a single state at a determinate moment of time. When the measurement intervenes in what had been only a probabilistic event, in which many states had been possible, the event becomes determinate, and only one state is found to be actual. The defender of branch attrition will naturally find this appealing. In contrast, the Many Worlds interpretation hypothesizes that not just one, but all of the possibilities in an indeterminate situation are actualized. For every possible state of a particle, there is a world in which that state is actualized. On this interpretation, that is what it means in the first place to say that many states are possible: Namely, that there are many worlds, each of which includes one of those states. Events are branch points into those many different actualizations, and at those branch points, the world branches into that many different worlds. The wave-function does not express statistical probabilities of a single-event occurring, it expresses the statistical distribution of multiple events actually occurring.

If this is true (I have no way to judge the physics), and each event generates alternative worlds, and so generates alternative time lines, then there will subsequently be many ways of dating the same event according to the different worlds' different time lines. Some versions of this theory work like Lewis' possible worlds, in that the alternate worlds do not interact once they are differentiated (although the same laws of physics hold in all worlds). In other versions, like that of David Deutsch, the different worlds interact.

The physicist David Deutsch[28] is a current defender of the Many Worlds interpretation developed by Hugh Everett[29] in the 1950s. His starting argument is drawn from the famous two-slit experiments (see the detailed descriptions at Deutsch *Fabric* 42–53). Under certain conditions, single photons shot through a slit at a screen will arrive in positions that make a distribution pattern that one would have expected if a stream of many photons had been emitted instead of one at a time. Deutch's argument is that for one photon to "know" where to go on the other side of the slit, in order to play its role in simulating a pattern that it does not individually have on its own, each photon has to move as if it were being affected by other possible photons, as if a lot of photons had been moving together at the same time. Each photon is just one of many that could have been emitted at that moment, but it operates as though all of the other possible photons were interacting with it. Therefore, Deutsch concludes, each individual photon is affected by "shadow" photons from other possible worlds. This is certainly odd.

Physicist Lisa Randall[30] describes the phenomena in the worlds that do not occur in our world, but interact with ours, as "virtual." "The net strength of an interaction is the sum of the contributions from all the possible paths that could occur" (Randall 230). Possible paths, although not followed by actual particles, are nevertheless "virtual particles" on "virtual paths," and these virtual particles all have actual effects on actual particles. The greater the distance that a particle moves, the more virtual particles interact with it (this point almost sounds like commonsense), and the influence of virtual particles on what actually happens is the "virtual correction" of the event.

For Deutsch, many phenomena in physics only make sense if multiple worlds interact. For example, one would expect charged particles in an atom to move out of position and crash. "It is only the strong quantum interference between the various paths taken by charged particles in parallel universes that prevents such catastrophes and makes solid matter possible" (Deutsch *Fabric* 213).

As is the case with David Lewis' possible worlds, Deutsch's different worlds may be so closely identical to my world that I cannot say which is mine. Yet unlike Lewis' worlds, Deutsch's other worlds contribute statistically to what happens in my world. What happens in many worlds round out the same statistical calculation. "All the environments I interact with are affected by what I do, and react back on me" (*Fabric* 308).

Other interpretations of quantum theory, in Deutsch's estimation, give too great a role to observers, as those interpretations need to explain how multiple possibilities get actualized in a single state, and they need observers to intervene to make that happen. (Observers do not have to be minds, of course, they just have to be physical interactions, but even that, he thinks, is too much for physics to take for granted every time anything is in a determinate state.) If the multiple (possible states) can stay multiple (by each getting expressed in its own world), and we do not need to end up with a single state, then we do not need observers to play a role in physics. In his view, assuming that observations determine reality is even more of a burden than assuming that reality consists of multiple universes. "Parallel universes are cheap on assumptions but expensive on universes" (*Interview* 84).

Where are the other universes ("universes" are the same as "worlds" in this context)? "In a sense, they are here sharing the same space and time with us", although we can only detect them indirectly (*Interview* 83–4).

The collapsing probability wave-function interpretation of quantum physics often uses the term "decision" to refer to the moment when a probability wave reduces to a determinate event or converges on a single world. The term "decision" is more or less equivalent to an "observation" that measures a single location or velocity. Finding Schrödinger's cat either alive or dead could equally be called an observation or a decision. In the terms of Deutsch's interpretation, to understand the meaning of a decision, we would have to follow it through the superimposition of multiple worlds, and this is what makes it interesting for our theory of a decision's multiple virtual futures. Only a decision distributed over multiple worlds is a genuine decision. To try to see decision in a single world would be to represent it as a merely a two-dimensional snapshot of a decision, whose simplified determination contradicts the multiple event it is supposed to represent. A single time in a single world, represented as a snapshot state of the world, is not a scientific description of anything. However, as we will see, the odd conception of time that follows from Deutsch's account has the by-product of reducing worlds to just these sorts of snapshots.

For Deutsch, the temporal implication of multiple universes is that across the multiverse, there is no sense of "at the same time" (*Fabric* 278; this is of course also a central point in special relativity). Since on his view, every moment of time branches into, or just is, different worlds, then every time there is a different time, it takes place in a different world. To speak of a different time is nothing other than to speak of a different world, and vice versa. There is no sense in which there can be "different times" within the same world. Deutsch drives himself to say that each world by itself is the length of a "snapshot." "Other times are just special cases of other universes" (*Fabric* 278). This he calls "the quantum concept of time." This is extremely odd. But is the distribution of times into different worlds so different from the division of a single world into different times, especially if the single world contains possible futures?

Deutsch is prepared to draw many odd consequences. Since our world's past is not really a different time in our world, but a different world that we call our past, any given alternative world might in principle count as the past of our world How were we ever able to "discover" the worlds we call our past? Only because our snapshot-of-a-world contains "more evidence" of worlds similar enough to ours to look like our past, than it contains of all the other worlds we can imagine. Presumably, this is also how we discover the worlds that we count as our future.

Some aspects of this account do not seem promising for our purposes. Understanding the future of decisions does not really proceed by finding evidence that certain other worlds are similar to our own. But the interesting thing for us about Deutsch's Many Worlds model is that the worlds fit together like a jigsaw, so that one universe's events are constructed out of interference patterns based on what happens in all the worlds together. Deutsch's branches do not undergo attrition as futures become actual.

> There is an infinite number of [universes] and the number is constant; that is, there is always the same number of universes. [This does not seem to me to be what Deutsch should say here.] Before a choice or *decision* [my italics] is made, in which more than one outcome is possible, all the universes are identical, but when the choice is made, they partition themselves into two groups, and in one group one outcome happens and in the other group another outcome happens. Normally these two groups do not affect each other thenceforward, but as I have said, occasionally they do. (Deutsch *Interview* 85)

Deutsch himself resists the branching metaphor found in Everett's version of the Many Worlds interpretation: "When I say the universes partition themselves into two groups, Everett said that one universe splits into two universes" (*Interview* 85). Still, with this proviso, Deutsch's worlds are like branches in that they spell out multiple futures.

Deutsch is willing to speculate about time travel, although with the proviso that the so-called past is really a different universe. If I travel backward in time, what I am really doing is traveling to a different universe similar to mine. When I return from my time travels, I am not necessarily returning to the same world I left; I might be returning to a different but almost identical "counterpart" world (*Fabric* 305). If the latter, then there will be no time travel paradoxes. Most time travel scenarios have the paradox that if I go back to the past and kill my younger self, that means that I will never have arrived at the present time in order to go back into the past to kill anyone. But the Many Worlds model of time travel would mean that when I go back to the past and see myself, I am really going to a different world, and seeing a different one of myself. I may have departed from one world, arrived at a second, and returned to a third. Indeed, if it takes time to travel through time (a sticky point in time travel stories), then I would get back to my own world at a different time than I left it, and that means necessarily that I return to a different world than I left, because no world contains more than one time. In any case, killing my past self does not kill my present self, since the different versions of me live on different branches, that is, in many different worlds.

In short, following Nahin's distinction, I cannot change my past along my own time line, but I can affect a different possible world, the world of my so-called past. Strictly speaking, "changing the past" does not change any world, it rather decides which world—which of the many branched worlds—to be in (*Fabric* 309). Conversely, visitors from the future could tell us what happened in their branching world (assuming they know which worlds count as their past), but they cannot know what happens in *our* future (i.e., which worlds we will be counted as the past *of*) (*Fabric* 313).

Deutsch's conflation of different worlds and different times cannot ultimately be good for a theory of time. Past and future may in a sense be other worlds, but they are different from other current worlds. The price that Deutsch pays (he would not have had to pay this price had his theory of time not got sidetracked from the main branch of his Many Worlds theory) is that on his description, each universe has no complexity of its own, and in particular, no temporal complexity, no duration, no delayed effects, no projects, and no futural decision contents. "Physical reality is the set of all the universes evolving together, like a machine in which some cogwheels are connected to other cogwheels; you cannot move one without moving the others. So the parallel universes are connected as inextricably as the universes of the past and the future" (*Interview*

89). Deutsch ultimately has something of a block universe, rather than a branching universe, in mind. As multiple as they are, the worlds he posits are all part of the unalterable machine, rather than expressions of creative indeterminism.

Deutsch's idea that probabilistic phenomena should be interpreted as a plurality of potentially interacting worlds, whether it survives in the physics of the future or not, is a helpful motif for a philosophy of decisional futures. But if we want to focus on branching events not as cogs in a machine, but in their undecided state, we will have to think of branches as undecided too. In fact, we can glimpse such a situation from the way Nobel Prize-winning physicist Murray Gell-Mann[31] describes "decoherence."

For Gell-Mann, it is best to picture the branches of quantum events over time as branches shifting around as they are blown in the wind. The branching histories of particles must be as indeterminate as the particles themselves. Otherwise, branches will sound like determinate states entailed by a common antecedent; events will sound like classical phenomena rather than the quantum phenomena we know them to be. If branches were determinate, then the universe would not really "form a branching tree." Since the quantum universe in reality contains interfering histories of quantum states, it does not allow any particular result even to be assigned a "true probability," let alone to be mapped as a series of world states. Probabilities exist only for "course-grained histories," that is, when real world differentiations are "summed over," that is, when we disregard all the different entangled probability states of particles. It is only when the differences are ignored, when they "sum to zero," when their entanglement collapses or is disregarded, when they are rendered "decoherent," that individual events can be assigned determinate probabilities and branching trees of events can be unambiguously diagrammed (Gell-Mann 145–50). In short, solid branches represent events which have been artificially reduced to determinacy by rendering their internal differences decoherent. To describe branches accurately, coherently, we have to sacrifice event determinacy, and allow branches to vibrate.

Commonsense thinks of branches as all-or-nothing disjunctions. But it is not only in quantum mechanics that we can think of a plurality of exclusive branches all followed together according to different degrees or different probabilities. It is not too hard to think of a person's decision leading her to follow several noncoinciding paths. One might say that a person follows one temporal branch at a rate of roughly sixty percent and the other at a rate of roughly forty percent. (Of course, each percentage point has to have its own degree of ambiguity, to keep the method consistent.) A single person, following a decision that itself might be made only probabilistically, or made in different but statistically interacting worlds, might be following a life path that could be described as sixty percent married and forty percent divorced, or sixty percent in military service and forty percent in protest against the military. In some cases, two patterns can simply be part of one consistent description of following a single branch. But in other cases, the descriptions may be incompatible; and yet even then, it may still be consistent to say that the person is following two different branches to different degrees over the same period of time.

In summary, the indeterminacy and truth-value gaps in future tense logic play out as probabilistic branches on the time lines of multiple worlds. How can we draw this back to a theory of decisional futures?

If a person follows many branches of a decision at once, is that the epitome of decision-making? Or is it more a case of indecision? Or is there a point when indecision and the best decision are the same thing?

CHAPTER EIGHT

Hegel: Morality without decision. Derrida: Indecision Theory

There is a fine line between the three theses: (a) it is often (or always) reasonable to make decisions in different ways, each of which would lead to a different possible future; (b) we often make more than one decision at a time, and thereby launch ourselves into a plurality of actual futures; and (c) we are often indecisive about which branches to follow, and thereby aim at the future ambiguously. In other words, (a) we can make many decisions about the future; (b) we do make many decisions about the future; and (c) we cannot be decisive about the future. Some variations on indecisiveness effectively function as future-plural decisiveness.

In fact, there is a fourth thesis, namely that philosophy of mind and action does not need a theory of decision at all in order to describe what we do and think about the future. Although we find terms like *proairesis* through the history of philosophy, and we might translate some of them as "decision," it is difficult to find a thoroughgoing theory of decision and the decisional future before the twentieth century. This may be an oversight, or it may be a conscious judgment that decision is not a very important notion.

I will discuss a sample philosophy of action that assigns only a minimal role to decision, namely Hegel's phenomenology of spirit. Hegel does say a few things about why decision is not a great category for describing the mind and the world. Hegel is typical in modern philosophy for underplaying decision. But Hegel is also a special case, in that his dialectic effectively transforms the flaws in decisional thinking into another kind of thinking that does a better job at decision-making than decision itself does. As we look at the "Morality" chapter of the *Phenomenology of Spirit*,[1] we will see connections among all four theses of decision: The plurality of decisional judgements, the plurality of decisional futures, the advantages and disadvantages of indecisiveness, and the effacement of the category of decision.

Hegel's critique of decision

There is a natural explanation for why Hegel might not find an important place for decision in phenomenology. The very idea of deciding among alternatives, discarding available pathways, does not fit well with an all-inclusive dialectic. Obviously, decision is a phenomenon of consciousness, so Hegel has to account for it somewhere. But it may be a deceptive form of consciousness, for one of the three reasons. First, it may be that only trivial issues divide into exclusive alternatives requiring decision. Important issues, like consciousness becoming self-conscious, may have only one way to go. Second, different decisions might be important in some contexts, but only temporarily. Different people side with faith or enlightenment, for example, but they all converge later when the terror hits. Third, alternatives may be false antinomies, in which case it would be best to synthesize instead of deciding among them. Hegel says a few words about decision (*Entscheidung*) in morality, although generally what he says is dismissive. And of all the things Hegel says decisions might be about, the future is not one of them. My goal is to see if there are resources in Hegel's text that might give decision a more interesting and more futural role than he gives it.

Hegel's Morality chapter begins with the connection between the willed world and the actual world. The contradiction, Hegel says, thinking of Kant, is that on the one hand, duty must act on an indifferent nature, and so cannot expect to change the world or result in happiness; but on the other hand, it is the "task" of morality to determine the world. The result is projection into a "future infinitely delayed" (This is Miller's translation; the German, *Unendliche hinauszuschieben* does not mention the future) (Hegel section 603 368/G446).

As Kant separates the spheres of natural and moral phenomena, he cannot allow moral action to have a causal effect on the actual world, only on a supposed world (section 604). The moral will is impotent for Kant. For Kant, that is fine, because that is just what morality is. Indeed, Hegel does not aim to abolish Kant's distinction between will and world. He just aims to break the will into finer grain, so as to mediate the gap into mini-gaps.

The fault line into which Hegel breaks will down into specifics is the fracture of duty into many duties. And the first job of the split into many duties is to make the will more concrete, so as loosely, although still inconclusively, to connect it with reality. But as we will see, the split into many duties creates problems.

This antithesis of one versus many duties (section 605) is largely the motor of Hegel's Morality chapter. To start with an experience everyone is supposed to recognize, a moral subject has one sense of "pure" duty, but also recognizes "*many* duties," whose principles and goals, Hegel claims, necessarily conflict. Hegel does not give an example of many duties here, but here is an example of my own: I have a duty to be honest, but I also have a duty to make other people feel good about themselves. That is, I have a duty to be honest, and a duty to respect others. But to respect others, I have to be a bit dishonest about what they really are. In short, I have one duty; correlated with that duty, I have many duties to apply it in different ways; the many duties can be inconsistent with each other; and the many duties can be inconsistent with the one duty I started with. Whether or not this is a good example, it seems uncontroversial that we have many duties. From this seemingly obvious point, Hegel draws three surprising inferences. Gradually, the divisions introduced in these inferences will constitute a kind of divided self that takes itself to need to make decisions.

First, the many particular duties take second place behind the one pure duty: They "have nothing sacred about them" (section 605). In my personal experience, I have to say, I experience conflicting duties, but I do not really experience having one "pure duty". If the many duties were in fact more sacred than the allegedly one pure duty, as it seems in my experience, it might throw Hegel's argument off. But it could easily just be that there is something morally wrong with me.

Second, if moral consciousness is primarily the overall sense of duty, the many duties do not exist for *that* same moral consciousness. The *many* duties "exist in *another* consciousness" (section 605). Hegel says this many times. Split duties imply split consciousness. Hegel's picture is not entirely clear. Perhaps the idea is that pure duty belongs to one consciousness and the many duties all together belong to another consciousness, as if a moral agent has an ideal self and a practical self. Or perhaps each one of the many duties has its own consciousness, as if one becomes a different self with each moral decision. After all, Hegel says, "in the actual 'doing,' consciousness behaves as this particular self, as completely individual" (section 606). If all moral consciousness ever did was to *contemplate* duties, the same consciousness could contemplate many of them. But because moral consciousness includes action, moral activity is individuated as a different consciousness each time it takes up a different duty.

The third inference from the split into many duties is that the place where the many duties are actualized is not quite the actual world. "Consciousness pictures its [moral] content to itself as ... a being existing only in thought" (section 612). Normally, for Hegel, picturing (*vorstellen*; on the next page, Hegel will turn morality into *verstellen*, a mockery of itself) is a symptom of a degenerate object. But picturing is the appropriate type of cognition for moral consciousness. A moral object is precisely not a natural fact, but a putative replacement. It is because of the trajectory from duty, to many duties, to many versions of the self, to many versions of the world, that the world for moral action is virtual. As Hegel puts it, "there is no moral existence in reality (*wirkliches*)" (section 613).

Consciousness is well aware of its own "insincere shuffling" between is and ought (section 617), so the pretense of moral action is in a sense successful. Morality may be overly idealistic, but it knows that it is. It is not that if we believe we are good or happy then we will be, and it is not that if we strive for ideals then we improve. It is that moral action is achieved on a kind of reality that "was never supposed to exist in the present" (section 618). In response to Kant's "postulate" of a good result that we know cannot be achieved at any time, Hegel's plan is not to overcome the antinomy with a synthesis that claims that morality can be achieved. In Hegel's morality, "the lack of fitness between purpose and morality is not taken seriously (*Ernst*) at all" (section 619). We do not do morality with a straight face.

Of course, this is full of twists. "Consciousness must go still further in its contradictory movement, and of necessity again dissemble this suppression of moral action" (section 622). This reads like Adorno. Morality dissembles its duties by equivocating about its actualization; then it dissembles the dissembling, and by redefining equivocation as success, it thereby succeeds. This is close to hypocrisy, indeed Hegel insists on this, because it is conscious that it does not take its duties seriously ... but not quite, since it takes responsibility for its con. Moral consciousness knows it is not worthy, but wants to be happy anyhow, and so it admits to wanting happiness without having to be moral, and accepts that when an immoral person flourishes, that is "the way of the world." This is not to scorn (*verschmäht*) morality, but to save it from reality; the joke is to save the one duty from the many.

Now, insofar as the self still attributes morality to itself, in the face of all this, Hegel calls it "conscience": "the third[2] self" (section 633). The first self is one; the second self is plural; the third self is conscience. The self is in good conscience if it can see its plurality as unity. The price is that conscience counts itself moral only by giving the last word to subjectivity. But this means that conscience attributes value to how we feel about our actions, devaluing action per se. The way it suspends the value of action, is to split action up into a plurality, and then to reject anything that is mere plurality. At first, the division of duties was meant to connect duty with reality by applying morality to different concrete cases. But now the division of duties is criticized for dispersing the self in realistic goals that it cannot really accomplish: Conscience distances itself from the many duties, and so justifies the moral retreat from reality.

The desire to avoid hypocrisy is the motive for not dividing duty into many. But Hegel gives an independent argument to show that even if it were desirable, there is no way to divide duty into many. Hegel's argument is that the division of duties leads to contradictions, and not the good dialectical kind.

> Conscience does not split up the circumstances of the case into a variety of duties... wherein the many duties would acquire, each for itself, a fixed substantial nature. If it did, then *either* no action

could take place at all, because each concrete case involves an antithesis ... a clash of duties—and therefore by the very nature of action one side would be injured, one duty violated; *or else,* if action did take place, there would be an actual violation of one of the conflicting duties ... The concrete shape of the deed *may* be analyzed by consciousness looking for distinctions into various properties, i.e. into various moral relations ..., or else compared and tested. [But] In the simple moral action of conscience, duties are lumped together [*verschüttet*: This could alternatively be translated as "spill over," "submerged," or "buried alive"] in such a way that all these single entities are straightway *demolished*, and the sifting of them in ... conscience to ascertain what our duty is, simply does not take place. (Hegel section 635)

In short, if there were a plurality of duties, then either some of them would violate the others in practice, or else nobody would ever do anything in practice. If we tried to accomplish distinct duties, eventually we would contradict ourselves. And if a single duty had genuinely different ways of being accomplished, then there would not be a consistent principle it was telling us to follow. There cannot be a plurality of duties; there cannot be a plurality within a pure duty; and there cannot really even be a gap between pure duty and the many duties.

I will try to give plausible examples later, but I am not sure that it would be immoral or impractical to hold oneself to inconsistent projects of the sorts Hegel critiques. I will consider below the moral theory that vacillating decision-makers are often morally the best. But for Hegel, a moral person should not separate possible good actions, and then compare them to test or judge which is better. A moral person should lump them all together and demolish the whole idea of moral judgment. In any case, such judgment, he says, simply does not take place. Hegel's assessment is certainly radical: There are no moral options to compare, and nobody has ever even tried it.

Now, read in a certain way, Hegel's rejection of any option selected out of a fluctuating set, is not so different from affirming generalized fluctuation. Not being committed to any single option is almost like committing to all options. And perhaps this would explain the crucial next move that Hegel makes so quickly, namely from a self with conscience to a self "recognized and acknowledged by others" (section 640). This move from self-reflection to mutual recognition is of course typical for Hegel, but how do we get to mutual recognition from the conscience who does not decide? Maybe this way: Lumping together unmade decisions constitutes what Hegel calls the "common element of the two self-consciousnesses" (section 640). I lump my options together, therefore my options are the same as everyone's. Perhaps this is less a rejection of many duties, as a distribution of many duties into many other consciousnesses. If I personally cannot unify morality into a single action with a single future, perhaps the community of moral agents as a whole can do it.

It is at this point that Hegel names decision: The point at which a consciousness is split by duties, recognizes the plurality of duties in the plurality of other people, but is still trying to contain that plurality in its own single consciousness, by evaluating all the options in its own single, conscientious act. The conscientious knower tries to "take all the circumstances into consideration" (section 642). It is really too late to take this seriously, as it already knows that it does not know all the circumstances that other people know. But this is one of the rare passages where Hegel uses the phrase "to choose between duties and make a decision" (*unter ihnen zu wählen und zu entscheiden*). Making a decision presupposes a world broken into parts—not only broken into alternatives and into a plurality of intersubjective agents, but also broken temporally into past, present, and future—where:

> reality is a plurality (*Vielheit*) of circumstances which breaks up and spreads out endlessly in all directions, backwards (*rückwärts*) into their conditions, sideways (*seitwärts*) into their connections, and forwards (*vorwärts*) into their consequences (*Folgen*). (Hegel section 642)

This is about as far as Hegel ever goes in the direction of moral consequentialism, and it is almost the only passage in Hegel where morality is directed toward the future at all.

As if in distaste at even mentioning *Entscheidung*, Hegel directly moves to the stage when consciousness admits in retrospect its bad judgment at having made a judgment. Its knowledge of the relevant circumstances was inevitably incomplete: "the pretense of consciously weighing all the circumstances is vain" (section 642). On this point, Hegel is certainly right. Admittedly, a person has to make a decision even when based on incomplete knowledge—Decision Theory is after all mostly about making decisions "under uncertainty." So paradoxically, Hegel says, as incompleteness is inevitable, and a person can only do what they know how to do, "[every decision] has to be sufficient and complete just because it is its own knowledge" (section 642). Of course, it is not strictly the case that decision-makers use all the knowledge they have access to—in reality, we generally use partial heuristics—but it is true that what we are conscious of is the same as what our decision is. So while a decision is for one alternative, the decision is not really an alternative to a different decision. That is, the option has an alternative (which could be stated in a proposition), but the decision has no alternative (at least, none that is as powerful for that consciousness). The selection of an alternative is consciousness's one and only and whole decision. Hegel puts his conclusion in a surprising way: The more knowledge a moral decision uses to select its alternative, the more it is a case of "arbitrary will" (*Willkür*, section 643).

A parallel passage from the *Philosophy of Right*[3] makes the same point. "A decision (*Entscheidung*) in favor of one or the other position depends on subjective arbitrariness (*subjektive Willkür*)" (*PR* section18, 28/69). Arbitrariness does not mean flipping a coin, and it does not mean distorting reason with wishful thinking. "Arbitrary" for Hegel means just what Decision Theory calls "rationality": "Using intelligence to calculate which impulse will give most satisfaction." By disparaging calculation, Hegel is offering a funny rule: Do what is right but do not try to decide what is best.

Admittedly, it is odd to say that calculation is arbitrary. But the support for this conclusion that mostly everyone knows, is that collecting, testing, and calculating evidence does not provide a common "measure" (*Mass*) by which to compare substantially different options. Hegel's favorite example of correct but biased calculation is drawn from criminal justice reform: Those who want criminals to pay compensation measure the cost of crime; those who want criminals rehabilitated measure the human capacity for reform. Objective calculation presupposes a qualitative decision previously made subjectively. In other texts, Hegel gives more examples of different people legitimately calling different things duties: Some people think property is good for family values, some think property is violence; some people think courage is a duty, others find it antisocial.[4]

The person choosing the opposite of what you choose always has a good reason, and, Hegel warns you in advance, "there is no use objecting" to that fact (section 644–5). The moment a decision is determinate, "it stands on the same level as any other" (section 645). And there is no point siding with whichever duty is more universal in its impact, because there is no ultimate reason to value the common good over the individual. "Calculating" duties is not just vulgar; it collapses under the problem of measure.

But the problem is not only with calculating the best direction. The bigger problem is that calculation aims at selecting "just one direction" (*einfache Richtung*). There is no calculus able to assign many different directions top priority at the same time. That, after all, is why this is important for a philosophy of right, and especially Hegel's dialectical philosophy of right: Decision is a bad paradigm for a state that needs to serve a plurality of interests.

Overall, then, the way to serve plurality all at once is not to split the future into a plurality of exclusive alternatives. And up until now, Hegel's argument is that ultimately there is no authentic plurality of options. But there is one point in Hegel's text where it becomes too late to deny differences among alternatives, and that is in the future. As long as I am in the process of deciding, in the present, using all and only the evidence available to me, there is just one decision to make, because my decision is essentially just my identity. However, "Once carried out" (*vollbracht*) then the decision has appeared in the "medium of being," in actuality, where duty is no longer able to "sublate (*aufhebt*) the difference" (section 648). It is not in their present, but in their futures, that decisions are different from one another, that their specific consequences cannot be ignored, and we are forced to second guess our previous convictions (section 648). The time when you cannot deny the role of decision, is after the fact. Hegel is anti-consequentialist, like Kant, insofar as the will is inner rather than outer; but Hegel is a consequentialist insofar as the will is future rather than present.

Now we should be getting somewhere. Decisions, where duties are many, are not distinguished by the evidence that preexists the choice, as all decisions make use of all the evidence; nor by different principles, since the good cannot lead to inconsistency; nor by the moment of decision, since all decisions carry the same conviction more or less; but by their futures. There is no real plurality of duties in the past or the present, only in the future. If we add that each moment of the future will contain further decisions, branching sequentially into layers more or less transparent to each other, we could imagine an interesting account of morality and time based on the historical unfolding of a decision. Not only that, but this might lead to the transition Hegel wants next, namely to forgiveness. The future is where other people are affected.

But this is not where Hegel goes with decision, it is just where I would wedge my topic of decision and the future into Hegel's dialectic. Hegel does not follow-up on the topic of carrying out decisions, and does not come back to the plurality of duties through the plurality of futures. To be honest, Hegel does not refer to the future at all in this chapter. My only hinges to the future were the three phrases I cited: The "delay (*hinauszuschieben*)" (section 603), the move "forwards" (*vorwärts*) (section 642), and the status of "once accomplished" (*vollbracht*) (section 648).

Instead of using temporality to get to intersubjectivity, Hegel appeals to language (*Sprache*, section 652) as the common ground of moral subjects, but I am not going to follow that thread here. Nor will I work out the last sections of the Morality chapter, where moral subjects try every conceivable variation on unmasking their own and each others' hypocrisy, dragging their heroes through the mud, accusing each other of posing as a "beautiful soul," and indulging in hard-hearted contempt for the other's morality, both whether the other confesses or not. The whole sphere of morality—I think this is Hegel's conclusion—is a morass of mutual contempt. The lone sentence about forgiveness in section 670 is not, it seems to me, the beginning of a moral world based on forgiveness, but the moment where subjects forgive each other for being moral in the first place, and for ever having made a decision. "The forgiveness which [consciousness] extends to the other is the renunciation of itself" (section 670). In any case, for Hegel, forgiveness submerges morality in religion. To quote the last paragraph in the Morality chapter, "The reconciling *Yea*, in which the two 'I's let go their antithetical *existence* ... : is God manifested in the midst of those who know themselves in the form of pure knowledge" (section 671). Religion (before it in turn becomes philosophy, of course) offers resources for general reconciliation, but it does so at the expense of morality, and certainly at the expense of decision, by picturing a community with gods.

But before we leave Hegel's theory of morality, I return briefly to the three problems with decision, to see if there are further resources in Hegel where we might be able to find reconciliation within decision itself. The problem with decisions among alternatives was that either (a) their differences are trivial; or (b) they eventually converge; or (c) their selection is arbitrary. Here are three scenarios of my own to explore these problems.

1 Euthyphro prosecutes his father for his crimes. Of course, his decision to do so makes a difference for his future and his father's. But his name becomes synonymous with the critique of that decision, and in the long run, the particulars of his decision are effaced by the broader issue of the foundations of morality in general. One person's decision for one thing leads to other people making the opposite decision, and finally the concrete choices are less important than the ideal principle.

2 An electorate of a certain nation votes for a party that supports public health care. Of course it makes a difference. But parties come and go, and over time all the nation's decisions in this policy area regress to the mean.

3 Sometimes people decide to forego something (a career, an adventure, or a topic), and later, having a chance to pick it up again, have the feeling of picking up where they left off, that they had been leading that other life under the surface all along. Maybe—how would we prove it?—we all live out all our possible decisions, not just the ones we actually make. In practical life, obviously there is a limit to how much an unselected option is retained in actuality. But in moral life, decisions were only made for virtual reality anyways. So perhaps it is especially in morality that decisions have the one–many dialectic, where decisions select not one duty and not many duties but one and many duties in different virtual futures that exist simultaneously. Moral consciousness needs to make many decisions to get to this plurality over time, but once accomplished, the plurality is transformed into coexistence—not by de-pluralizing the decisions but by pluralizing the future.

So the specific way in which Hegel removes decision-making from morality, namely by dissipating the primacy of one alternative over another, and distributing the plurality over a field of futures, might give the moral sphere a last chance at reconciling moral decisions despite their plurality, before handing reconciliation off to religion. Not forever, probably. Hegel might well be right in the end that morality is a bankrupt form of consciousness and that ethics is the most useless part of philosophy. Still—and maybe this is just a small difference—instead of thinking that pure duty is the essence of morality, and that decisions among options are trivial, we might think instead that there are essentially many duties, so the fact that decision-making throws away the moral will in bits and pieces is morality's finest moment.

In other words, the very reason for thinking that decision is not essential to morality, that right is not decided, that plurality undermines the ideal of the best decision, is also a reason to formulate a new definition of decision that is essentially plural, where the plural decision is not a challenge to decision but is the best form of decision, where nondecision is the highest ideal of decision.

Every topic in Hegel ought to be historical, so why not read decision as a distinctive dialectic of future-directed consciousness? Then donate this dialectic to Decision Theory.

Indecision theory

The moral contrast decisive–indecisive is often associated with active–passive, autonomy–heteronomy, deliberate–habitual, purposeful–indifferent, success–despair, procedure–drift, strength–weakness, science–play, and the like. In classical terms, the first element in each pair is held to be morally preferable. In heretical terms, common today, the second seems preferable.

I will avoid any simple claim that being decisive is either better, or worse, than being indecisive. If one is watching a person being beaten, a decisive intervention may be good; if one is asked to shun a family member, indecisive delay may be good. Rather than decide (or not) when it is good to decide or not, it may be that the best thing philosophy can do is just to analyze the temporal structure of moral decision making, to develop techniques for explicating the layers, periodicities, and long- and short-term extensions of decisions, to map the threads and options in each decision's temporal projections, and to make explicit the non-synchronicities and the inconsistent degrees of projection in decisions where they exist. Even decisions that end in, or at least pass through, a determinate state, like the decisions analyzed by Karen Houle regarding abortion, may be so filled from beginning to end with ambiguities and multi-rayed directions, that the case that will be decided should nevertheless be described "sexuate undecidability."[5] Such an approach toward undecidable decision-making might make headway in the descriptive phenomenology of morality, although it may not solve much normatively.

To give a quick example of layered temporal description, consider the moral dilemma in the film *Eye in the Sky* (Gavin Hood, 2015). The plot is that military and political leaders in the United Kingdom, the United States, and Kenya are deciding whether to send a drone missile to destroy a group of terrorists presumably on their way to a suicide bombing. The problem is that a girl is selling bread beside the target zone and is likely to be killed by the missile. Although the film seems to favor the drone attack, it leaves ambiguities about which decision is the best. The action is slow enough to encourage debate among filmgoers. But the point on which the film is decided without ambiguity is that decisive action one way or another is better than indecision. The characters on both sides of the question who are portrayed as wise, professional, thoughtful, and experienced object to the indecisiveness of other characters on both sides who are portrayed as being afraid of their bosses, public opinion, litigation, and even their own bodies. The limited exception is that the human pilots operating the drones by remote control are portrayed positively, although they insist that the decision be delayed to protect the child if possible. Still, when the order comes to release the missile, they do so without further delay.

This is a lifeboat tale: A definite child's very likely immanent death valued against a large number of indefinite people's possible deaths that would follow from various actions the terrorists are predicted to carry out with various degrees of probability. (The guilt of the terrorists is never in doubt; their motives do not come up; it is a given that they deserve to die, and that the joint military has a moral right to kill them.) I do not know what I would have done. But if we want to analyze the temporal structure of the decision in the film, there are a few points to consider. For one thing, only military and political authorities are consulted; taking more opinions of regular people would take more time. Decisiveness certainly narrows the scope of democratic involvement, as it has since ancient Athens. Most important for our method of disentangling the temporal projections within a decision, the decision-makers use different methods to consider the future branches of the two options: Killing the terrorists and killing the bystander. When forecasting branches around the decision to kill the terrorists, the characters consider all sorts of variants: Whether they could let the terrorists leave the scene where the child is, and still stop them before they reach the targets they plan to blow up; whether public opinion will side with the military under various scenarios; whether some terrorists might escape the bombing; what message various results might have on terrorists elsewhere; and so on. Obviously, in an action film, not every real world consideration will get expressed, but the movie largely consists of characters considering their options over both short- and long-term. However, when the characters are considering the topic of risking the child's life, they do not look past the moment at which the child will either be hurt or not. They do try to get the child away from the danger zone, without alerting the terrorists. [SPOILER ALERT] But when the child is in fact injured by the drone missile, it becomes clear that the decision-makers had not considered any options like sending medical teams to the point of impact, or interacting with the neighboring community in the aftermath. In short, they project one option into the long-term, and the other only into the short-term. The temporal branching schemata they use in their decisions is inconsistent and nonsynchronous.

Who can say what difference it would make in the minds of politicians if they used more complex temporal projections in their decision contents, if they noticed their temporal options, and not just their actional options.

The classical opinion that decisiveness is good and indecisiveness bad is still common. Hundreds of books and articles on decision include the phrase "paralyzed by indecision." (Google it.) This generally comes with the helpful advice that we should understand the cause of indecision so that we can move on. Indecision is associated with vagueness, neutrality, disagreement, fragmentation, cowardice, and brooding. Each of these associations would deserve analysis, but my idea that part of successful decision is indecision, or what I am calling the persistence of plurality in decision, associates indecision with none of the above.

Sometimes, it seems at least prudent to be indecisive. For example, for an investor, there are times when it is best to hang on to cash and decide later whether and where to invest it. There are times when it is best for a prisoner facing the prisoner's dilemma to stall for time. In some cases, it might be plausible to disavow responsibility for a decision made by a nation in one's name, especially if the disavowal takes place early enough to have the effect of weakening the decision. In some practical contexts, it is pointless to act decisively, because a decision is going to be second-guessed and counteracted no matter what. For example, the Visual Effects Supervisor Jamie Dixon writes: "One of the most difficult things about filmmaking is the decision-making process. And it almost doesn't matter how early you get started trying to solve a problem, because nobody's going to make up their mind until about the last six weeks of the show anyway."[6]

My thesis is that there is a genuine connection between non-individuating or inconclusive decision on the one hand, and indecision on the other.[7] But the choice of terms is tricky. On the one hand, if the term "indecision" is used to mean indifference, disinterest, docility, or thoughtlessness, as its detractors assume, then it would have nothing to do with affirming a decision with a plurality of alternative branches. In that case, it would simply be better for us to talk about plural decisions and not try to defend indecision. On the other hand, if the term "indecision" is used to value pluralism, that seems to assume that the term "decision" conversely means opting for one future to the exclusion of others. But this is the old definition of decision, which I am looking to reverse. I want to use the term "decision" to cover the same pluralism that I want to use the term "indecision" to cover.

To accomplish this, we could reserve the word "decision" precisely for cases where rules for processing evidence and predicting the future cannot determine what to do. To put it strongly, we could use the word "decision" just for judgments about situations that are undecidable, that is, just for those turning-points that go in many directions, just for those moves that are not completely compatible and not completely committed. In sum, if computation is deterministic and "decision" is voluntarist, then decision would share with indecisiveness the recognition that the grounds for conclusiveness are lacking. At some point, it is a matter of semantics whether we say that indecision is the opposite of decision, or that indecision is similar to decision insofar as both imply future pluralism.

Badiou is a good example of a philosopher who uses decision and undecidability almost as synonyms. In Badiou's terms, if all of a situation's elements are both apparent ("presented") and self-explanatory ("represented"), then the situation is normal and calls for no decision, either to interpret it or to intervene in it. But those situations that have an internal gap or unresolved conflict (an "evental site"), an impasse whose definition and potential resolution are undecidable, are just the situations that call for a decision (an event). In this rather odd sense, Badiou calls Hegel a philosopher of decision (and Kant a philosopher of judgment)[8].

If we were less interested in temporality, Badiou's account of decision in *Being and Event* would play a more central role. But his interest in decision is less about projected futures and more about ontological choices and interventions. Even his discussion of the subsequent fidelity to decisions largely treats time as a secondary function of such procedures as enquiry, verification, and forcing.[9] A more relevant proponent of indecision in the context of temporality is the philosopher and former mayor of Venice, Massimo Cacciari. Cacciari, whose politics are left and green, was called indecisive for his slow response to the need to protect Venice from falling into the sea, an application of Indecision Theory that may not show it to advantage. Still, there are many interesting features of indecision that arise from his work.[10]

Cacciari's case studies for his "phenomenology of decision" are drawn mostly from fictions: Hamlet, K (from Kafka's *The Castle*), and Godot. He suggests, for example, that Hamlet's (Hamlet is for many writers a poster child for indecision) attempt to decide whether to take revenge is itself what creates conflict and tragedy—if Hamlet had considered the possibility of forgiveness, which Cacciari treats as a kind of indecision, "redeeming" the past instead of trying to force a future, things would have turned out better (Carrera 127). It is not clear how to assess what might have happened in fictional stories if they had proceeded differently, but there is no doubt that this is the sort of value Carl Schmitt would abhor (126), particularly when Cacciari turns indecision into a value for real world politics. In the context of a multicultural Europe, where "deciding to do this and not that" implies, Cacciari thinks, where the domination of certain nations over others is the likely result of decision-making, "being undecided is not necessarily a bad thing for Europe" (135). Assuming we use the term "decision" to cover branching into pluralized situations, then the multicultural variations that Cacciari values will result as much from intensively complex decision as from indecision.

The Turing decision problem

Even in formal terms, in computational systems, there is already an undecidability problem. The problem of deciding when a solution to a problem has been reached, and of knowing generally whether a system is capable of completing a computation, is Turing's version of Hilbert's "Decision Problem" (*Entscheidungsproblem*).[11] What Turing proves is that certain computational problems are undecidable even for the best computation machines. It seems unlikely that the broad problems of decision that we are interested in, like how a person carries out a morally complex decision that involves many people at different times, are amenable to computation. Still, the formal structure of the decision problem for computational systems is interesting in its own right.

A Turing machine does nothing but calculate arithmetical solutions mechanically. The machine is programmed to progress back and forth by certain rules among a row of boxes, checking a 1 or 0 in each box as it goes, or switching the digit under programmed conditions. Depending on which box, with which digit written on it, the program makes the machine stop at, the result can be interpreted as the solution to an arithmetical question. If we program a machine to add 1 + 2, we need to know which box with which digit on it "means" 3, so that we can see if the program we wrote for the machine made it stop on that box. The machine obviously does not know that that box "means" 3, it only knows how to follow the instructions and when to stop. If we find that a machine does a lot of additions correctly, we might trust it to do additions in future.

As all calculations are digital in the same way, different arithmetical questions are different only because of the interpretation we put on the results; the program has the same small range of procedures in every case (go back or forward one box, write a 1 or 0 in a box, repeat or do not repeat, and things of that nature). In that sense there is only one Turing machine, a "universal" Turing machine. In a more restricted sense, we might say that each program that follows certain digital instructions is a different Turing machine. In either case, the Turing machine is the program, not necessarily a physically constructed machine. We evaluate a Turing machine by asking whether it will be able to solve a certain calculation problem.

One reason a Turing machine might not be able to solve a certain arithmetical problem is that it was programmed to make the wrong calculations. But another reason it might fail to arrive at a solution is that it might get caught in an infinitely repeating loop. In fact, it is easy to construct a program that repeats infinitely. If it gets into a repeating loop before it finishes its calculation, then it will not stop, or make a decision, or get to the solution. Some machines stop and reach a solution, and some never stop. Normally, it is a finite task to determine whether a given program will stop or loop.

The interesting problem is the meta-question. Can a single Turing machine be programmed to decide for *any and every* other Turing machine whether or not those machines will get caught in an infinitely repeating loop? And when the meta-machine starts working, can we know whether *it* will stop, or whether it will loop. Can we decide whether the machine programmed to decide whether every other machine decides, will come to a decision? To go still further, is the answer to this question computable; that is, can a Turing machine decide this? Can the Turing machine that decides whether any and every Turing machine gets into a loop, decide *about itself* whether it can arrive at that decision without getting into a loop? Can it decide whether it gets into a loop, without getting into a loop in the course of that decision? Can the universal Turing machine decide whether the universal Turing machine arrives at a decision? This is known as the Decision problem, or the Stopping problem.

Of course, the meta-machine does nothing but calculate in the same way as any other Turing machine: It goes back and forth between boxes, marking boxes with a 1 or 0, and so forth. The only way it can decide whether any given machine solves a problem, is to imitate exactly what that other machine does, with an extra step to determine whether that machine came to a final result. It is not difficult to design a machine to decide whether a particular other machine stops at a solution. But the meta-machine in question must be a "universal" machine in a strong sense, namely in the sense that its program can imitate every other machine. The problem is that this universal meta-machine has to run through every other machine's program before it can decide whether the other machines stop or not, and therefore before it can decide whether it itself is capable of evaluating any and all Turing machines. So while imitating one of the other machines, it might get caught imitating that other machine's infinitely repeating loop. No matter how many machines the universal machine has tested so far, it cannot decide whether it will get caught in one of the infinite loops of some other possible machine. So the universal Turing machine that tests Turing machines cannot test itself, that is, it cannot decide whether it can test itself without getting caught in its own loop while imitating a loop of some other machine. Therefore, the universal Turing machine cannot solve the universal decision problem.

There is a view that the paradoxes of Russell, Turing, and Gödel have direct consequences for systems of meaning, which help to explain and develop the philosophies of difference and multiplicity found in Derrida, Deleuze, Badiou, and others.[12] My own view is that these paradoxes apply only to axiomatized systems like formal logic and set theory, or to formal computational systems, and interesting as they are, have only a metaphorical relation to the sorts of issues I am dealing with here, involving decision-making and temporal phenomenology. Even so, from the field of computation we can keep in mind that decisions are never far from meta-decisions, so that even when a decision seems to have a conclusive result, the meta-decision as to whether decisions generally have conclusive results does not have a conclusive result.

Derrida on undecidability

For us, the most relevant analysis of decisional ambiguity and multiplicity is Derridian undecidability. The crucial point in Derrida, which some readers miss, is that undecidability is not the same as indeterminacy. Deconstruction is not relativism, or ironic play, or skepticism, or a hesitant type of indecision. Meanings are undecidable just because for any text, there is a plurality of possible future interpretations. Deciding on a meaning involves all sorts of temporal deferrals: Preimagining the future perfect, the time at which the meaning will have been decided, continuous and discontinuous patterns of transmission and reception, and so on. The points at which interpretations of a text branch are relatively determinate, although the turning points are themselves open to variation and further interpretation; the branches of interpretation are relatively distinct, although sometimes they converge later and their differences may then be erased; the history of changing interpretation over time can be traced, although interpretations are often premature or delayed; the plurality of interpretations are relatively describable, although the terms of their description open into further interpretations; different authors and commentators can be interpreted together, although intertextuality interferes with the identity of each.

Derrida is primarily concerned with the undecidable meaning of texts,[13] and less often with decisions to do something. Still, there is not a sharp border between judgements about meaning and decisions about future acts, and his discussions of decidability in meaning shape many of his concerns bearing on practical action. In a text on forgiveness, to take one case, Derrida treats the feedback between present and future, a typical hermeneutical problem in the reception of texts, as the source of the undecidability of moral decisions. For example, if a society decides to eliminate people with a certain point of view, later on few such people will be left to object, so it will look like the right and popular decision, but only because it erased the evidence that would otherwise have made it look like a bad decision. To be sure, this is not the only way that futures feed back on decisions. Sometimes the future clearly reveals problems with a past decision. Still, there might be some truth to the generalization that, "One is never sure of making the just choice; one never knows, one will never know with what is called knowledge. The future will give us no more knowledge, because it itself will have been determined by that choice."[14]

I will not rehearse all the arguments for multiple interpretability in Derrida's work—I only want to define undecidability. I draw from *Limited Inc* three reasons why meanings are undecidable.

First, reading a message is a way of repeating it, or reiterating it, or reopening it in a new context, and the difference between the context where the text was enunciated and the context where it is read makes it inevitable that something different enters into the meaning once it is read: This is the "law of undecidable contamination" (Derrida *Limited* 59). Of course, two contexts may be different on some parameters but not on others. Therefore, the first and second meanings of the text are not in their totalities either same or different. For that matter, each of the parameters used to measure sameness and difference may mean something different at the time of writing and at the time of reading. Therefore, we cannot even quite say that the two meanings are same and different in different senses: Each sameness and difference is itself subject to more than one sense. And further still, when the text is written, potential readers will already have been envisaged, so the second meaning, the reader's meaning, although largely still absent, is in some sense prior to the writer's meaning. Therefore, even the decision about which meaning is first and which second, about the time order of interpretations, and the decision about how open and/or prescribed the range of meanings is, are subject to a plurality of legitimate readings. The conclusion is that meanings cannot be "delimited" with "rigor."

The second general argument for undecidability in meaning is that because meaning (*vouloir dire* in French) begins with the motive of "wanting to say," the judgment of meaning appeals at some point to what was desired, and desire is always to some degree unconscious. As the unconscious is by definition unexpressed at any given time, all meaning intention includes some "undecidable" content (*Limited* 75).

The third argument is that because decidability presupposes a metaphysics of identity (*Limited* 93), all of Derrida's arguments against the metaphysics of presence (which I will not rehearse here) are also arguments for undecidability. Undecidability is virtually coextensive with many of Derrida's other terms, from "deconstruction" and "deferral," to "aporia" and "gift," to "hospitality" and the event "to-come."

Many of Derrida's most explicit comments on undecidability appear in "Afterward: Toward an Ethic of Discussion", the appendix to *Limited Inc.*[15] In this text, Derrida qualifies the limits of undecidability, aiming to show how it is neither relativistic nor hyperbolic. It is possible that in this text, Derrida understates the radical character of his position, but I think what he says here is right. He says that he has never said that undecidability was a matter of "complete free play" (*Limited* 115). Nothing is complete, not even incompleteness, not even deconstruction. Decidability is not an "all or nothing" question (*Limited* 117). The dream of unlimited undecidability appeals to the "pathos" of complete freedom, which is as one-sided as the presumption of unlimited presence. If there could be a text whose meaning were completely indeterminate, completely open to interpretation in any way whatsoever—that is probably not even possible—it would be senseless. It is not even conceivable that all texts could be completely indeterminate. Derrida says he has never done more than point out the "limits" of decidability in this or that particular context or text. It is this "trial" of decidability, the challenge to this or that authoritative decision, that makes decidability a moral and political topic (*Limited* 116). Undecidability is the name for probing the weak points in judgments about meanings, not a thesis declaring that there is no meaning.

Texts are always interpretable, and indeed, texts do not exist unless "interpretations start" from them (*Limited* 145). It is impossible for any text to prevent alternative interpretations, as much as its author might want it to, or as much as a text might seem uncontroversial on first reading. Although texts are unstable, it only means that there is a "relative indeterminacy" in the text. "It is not indeterminacy in itself, but the strictest possible determination of the figures of the play, of oscillation, of undecidability, which is to say, of the *différantial conditions of determinable history, etc ... .*" (*Limited* 145). If anything, an increase in undecidability is correlated with an increase in determinacy (*Limited* 147), as it is at a particular pressure point, developed through an intricate reading of the implications of the details of a text, that a divergence into different possible readings emerges. This is the key point: Meanings are *undecidable* on particular points that arise in the process of interpretation; meanings are *not indeterminate*.

> I do not believe I have ever spoken of "indeterminacy", whether in regard to "meaning" or anything else. Undecidability is something else again ... Undecidability is always a *determinate* oscillation between possibilities (for example of meaning, but also of acts). These possibilities are themselves highly *determined* in strictly *defined* situations (for example, discursive—syntactical or rhetorical—but also political, ethical, etc.) They are *pragmatically* determined ... I say "undecidability" rather than "indeterminacy" because I am interested more in relations

of force, in differences of force ... "Deconstruction" should never lead either to relativism or to any sort of indeterminism. (Derrida *Limited* 148)

In other words, undecidability arises not because there are no motives for deciding one way or the other, but when there are positive motives for deciding in more than one way, that is, where there are different "forces," and an overabundance of determinacy. A text does not have more than one meaning just because it is made up of words, it only has more than one meaning if there is evidence for more than one meaning. This is, after all, the sense in which undecidability includes difference. A point of undecidability is a point of divergent paths, each of whose points diverge in more than one way, that is, each of whose points diverge within themselves. The point of all philosophies of multiplicity, from Derrida to Deleuze to Irigaray to Foucault to Badiou, is that identities do not break down into smaller identities, but that each smaller identity breaks down (and builds up) into further pluralities, as far as can be (how far it goes is itself a branching possibility—it may go ad infinitum, but there is no rule that it always does). This diverging divergence is the "double bind" (*Limited* 148). Every turning point leading to multiple branches is already just one of its own multiple possible branches. A decision is not just whether to follow one branch in future or another: It is already also a decision to examine just those branches, as opposed to other branches; it is already a decision to decide on that topic and not another; it is already (as Heidegger said) a decision to make a decision as opposed to doing something else; it is already a decision to make the decision with or without certain other decision-makers; it is already a decision to make the decision at that moment and not at another; and hence it is already a decision to order time in a certain way.

In short, undecidability is all about the intertextual, intersubjective, and inter-temporal plurality of decision-making determinations. If interpretations and branching choices of action are undecidable, they are equally decided under those forms of plurality.[16] It is in these ways that deciding the meaning of a text, and all the undecidability that involves, applies also to decision regarding the direction of an act. I will take up just one application, from Derrida's *Politics of Friendship*.[17]

Decisions regarding friendship take time (Derrida is commenting on Aristotle), and cannot be confirmed quickly. This is a practical variant of the grammatological principle that signifiers can only be interpreted once they have been relayed, reiterated, and resubmitted to question some number of times in different communicative contexts. The affirmation that somebody is a friend, is a call for delayed decision-making: "The time of this decision in the ordeal of *what remains to be decided*" (*Friendship* 15). As in decisions about meaning, a decision in favor of action is defined not by what it settles, but by what it leaves still to be decided. That is why a decision takes place in two (or more) distinct times, yet two times forced into contemporaneous relevance, "two times in the same time" (*Friendship* 16). While one version of the decision is made, another remains undecided. In that way, there is a "perhaps" function attaching to every decision (*Friendship* 18). Or to put it in different terms, instead of saying that "perhaps" a decision has been made, we might say that a decision has definitely been made, but that there remains some "secrecy" over what it guarantees (*Friendship* 258).

Just as earnest decisions withhold secrets, dishonestly made decisions might eventually get acted upon anyways (*Friendship* 274–5). Authenticity is not a necessary condition for whether decisions have futures. Indeed, most, if not all, decisions are to some degree false, or at least half-hearted, finger-crossed, over-blown, or so-to-speak-decisions, so it is almost a foregone conclusion that it is the bad faith decision that we have a responsibility to carry out.

In the meanwhile, between the time the decision to be friends is set in motion and the time we can say it has been confirmed, there is plenty of time to calculate and reassess whether the other person is really a friend. In that meantime, deconstructive decision procedures may even take the time to apply Decision Theory procedures, measuring expectation values for different kinds of friends, balancing memories of good and bad times, calculating odds of betrayal, planning strategies for iterated mutual reaction, not to mention offering forgiveness for trembling hands. As Derrida puts it somewhat vaguely, there is a period of time when decisions submit to the "bond of time and number" (*Friendship* 20, 67). The timescale of decision-making with others varies, naturally: Making decisions with brothers, for example, generally carries more memory baggage than making decisions with friends (*Friendship* 294). But in every case, the decision saturates the time between question and answer with considerations on every side of the equation, with conclusions and second guesses (*Friendship* 250 and 254). At every moment, the decision arrives at some state or other, but then at the next moment, it does so again.

Of course, the decision to be friends is not up to one party. Acknowledging the undecidability of the decision is in part a way of passing responsibility to the other (and taking responsibility for giving responsibility to the other), whose decisions one cannot predict or control (*Friendship* 69). If the other is really unpredictable (Decision Theory does not think she is), then no matter how rational you are in your own part of the decision, the other acts as an exception to your rule, and makes your joint decision a mixed bag. It is as if Derrida wants to include friends, not just enemies, under the rubric of Schmitt's exceptional decisions. The other may be your friend, even your loved one, but the other is still the exception to your decision over time. For temporal reasons, "attraction leads to rupture" (*Friendship* 256). But the rupture in friendship is nothing but the irreducibility of the future, so as long as you love the future, you still love the friend whose friendship can only be decided in that future. Schmitt, of course, would never articulate it this way. He might conceivably agree with a statement Derrida attributes (without reference) to Kierkegaard, namely that "the instant of decision is madness" (*Gift of Death* 65[18]), but he would certainly not have agreed with the inference from the uncontrolled future of decision to responsibility for others as friendly amenders to one's own decision-making.

What is interesting in Derrida's schema for undecidable decisions is that ethical implications are grounded not just in shared subjectivity, but in the pluralized future. It is probably too strong to say that every decision with a certain temporal noema automatically has a friend as its object. Derrida finds room for resistance precisely in the way that even the strongest decision conceals at least some dissident responses and responsibilities set aside in the form of future possibilities (*Gift* 26). However, there is a connection between pluralistic decisional temporality and ethics.

The set-asides inherent in decisional futures show once again that decisions in plural time are decisions about virtual events in virtual time, not just about actual occurrences. When something remains to-come, it is on our heels like a "spectre" (*Friendship* 289).

CHAPTER NINE

Deleuze: Decision in the empty future. The virtual decision = X

I will use Deleuze to pull together elements for a theory of the multiple virtual futures of a decision. I use his discussions of future time in *Difference and Repetition*; of the open-ended object = X in *Difference and Repetition* and *Logic of Sense*; of decision as a dice throw (in my view, not a successful paradigm), from *Nietzsche and Philosophy*; and of incompossible worlds (I compare Deleuze with David Lewis on possible worlds) opened up by Adam's decision to sin, from *The Fold*.

Is the future full or empty?

To begin, we face a difficult problem both for the study of Deleuze and for the nature of future time. In *Difference and Repetition* (Chapter 2),[1] Deleuze defines the future as "the pure and empty form of time." How empty is it? In one sense, of course, the future has to be empty, since if it were full of events already, there would be no scope for making a decision about what to do in the future. But in another sense, if the only way to imagine the future were as an empty zone, then we could not imagine carrying out decisions in the future. What kind of status do these futures have, and in what sense are they empty? This is not just a technical problem in Deleuze studies. The future really does have this puzzling character. It is empty. Yet it is what concrete decisions are about.

There are not many places in Deleuze's corpus where he talks about the future. In *A Thousand Plateaus*, for example, all the chapter titles are dated with reference to some telling event, but none has a future date in its title. In the "Geophilosophy" chapter of *What is Philosophy?*,[2] Deleuze discusses the "new," which he associates with the "non-historical future": "Acting counter to the past, and therefore on the present, for the benefit, let us hope, of a future—but the future is not a historical future, not even a utopian history, it is the infinite Now… the Intensive or Untimely, not an instant but a becoming" (Deleuze WP 112; also 96 and 101). This passage names the future, but as I read it, it deals more with becoming than with the future. In *Cinema 1*,[3] Deleuze calls Buster Keaton's character "the man of the future" (*Cinema 1* 177), but this is because of Keaton's gags around newfangled machinery, not because his future is empty. Deleuze's only text focused directly on the future is Chapter 2 of *Difference and Repetition*, which develops his theory of the three syntheses of time.

I am not going to talk about the first two syntheses of time in Chapter 2 of *Difference and Repetition*: The syntheses of the present and the past. These are relatively short, and aside from the many points unresolved in the literature, we have a general idea about what Deleuze means by the expanding living present, and by the coexisting layers of the past.[4] In contrast, it is difficult even to formulate a general account of what Deleuze means by the third synthesis of time, or the future.

When Deleuze introduces the third synthesis of time, he does so to explain something about the past: How is it, if there are many layers of the past accessible in the present, that we can "live" in those layers, "explore," "search," "penetrate" (*DR* 85), "move" (*DR* 297), and "choose" (*DR* 83) among them? How do we decide which parts of the past to use in the future? Choice has something to do with desire, he says, and that has something to do with the future. But Deleuze treats many disparate issues under the heading of the third synthesis of time, from choice and decision, to Eros and Thanatos, to dark precursors and the object = X, to novelty and virtuality, to the form of time (Kant) and caesuras in time (Hölderlin). What exactly is the topic covered by "the future?"

Why is the exploration of the pure past erotic? Why is it that Eros holds both the secret of questions and answers, and the secret of an insistence in all our existence? Unless we have not yet found the last word, unless there is a third synthesis of time ... (Deleuze *DR* 85)

It is not hard to construe a simplified deduction to describe how the present (the first synthesis of time) implies the past (the second synthesis), and how the past implies the future (the third synthesis). First deduction: A moment is only present if it passes; the present cannot pass unless it has somewhere to pass into; therefore there must be some kind of virtual storage zone for times to go into; and once the past is all stored, it is all available for reuse at any time; this is what we call the past. Second deduction: If time is stored, it must be possible to explore or retrieve it; therefore there must be a kind of time in which such possibilities subsist; we call this kind of time the future. In short, storage implies future retrieval.

Obviously, the future has not happened yet. The future has no determinate content. But given that the future will make use of the past, is the future nevertheless full of determin*able* content? Does the concept of the future designate continuations, recurrences, reorganizations, recreations, and rethinkings of the present and the past, or does it do something entirely different?

Readers of Deleuze are familiar with his two senses of the past. In one sense, the past is just the set of previous present moments. But insofar as the content of a present is available for use at some later time, the present creates a past in a second sense, a past to live in again and again: What Deleuze (construing Bergson) calls the "pure past." In fact, a content of experience is already available for reuse as soon as it is present, so the "pure past" already exists at the same time as its content is present. This is the beginning of Deleuze's complex argument (which I will not rehearse here) to the effect that the past and present coexist. By analogy, Deleuze distinguishes two senses of the future: The set of present moments still to come, and the pure future. It is not easy to pin down the latter.

The trick for interpreting Deleuze is that there are three syntheses of time, which have the names present, past, and future. But each of these three syntheses organizes all of time in its own way. The present is the process of time passage, extending the present into past and future; the past is a system for the reactivation of past events in the present and the future; the future is a procedure for emptying events of their past meanings and projecting the unknown into the unknown. In this way, present, past, and future syntheses each describe movements that pass through all three dimensions of time—present, past, and future—according to their own definitions of those dimensions. The three *syntheses*, which are named "present, past, and future," each contain three *dimensions* of time, which unfortunately also have the names "present, past, and future." There are thus nine kinds of time, and they are all real. If we name each one as a coordinate pair combining a synthesis with one of its dimensions, the word "future" will appear in five of them: Future-present (the future dimension of the present synthesis—what common sense calls the future, the next present), future-past (the future dimension of the past synthesis—the past as it will be reactivated), present-future (the present dimension of the future synthesis—the present in which we desire a future), past-future (the past dimension of the future synthesis—the past that has a futural drive), and future-future (the future dimension of the future synthesis—the open virtual future, which Deleuze calls the pure future, and which we have yet to define).

Each of these senses of the future is empty in a different sense. The first sense of the term "future" consists of present times still to come. That sense of future is empty in an obvious sense, since the times to come have not yet come. In fact, it is so obvious that the future has not happened yet, that we might expect every account of time to say so. Yet in the first synthesis, the future is the expanded present; its not-yet is only a phase, eventually the future will arrive. And in the second synthesis, the events that fill the past, while of course not yet repeated, will survive as virtual paradigms into the future (the future-past). The future dimension within the synthesis of the future, the future-future, must be empty in a stronger sense.

Consider Deleuze's point that just because of time, events are determin*able*. Why would time alone make events determinable? Why does time not just measure already determined events at different times, including future times. How can what comes before fail to determine what comes after? Why can we not predict deterministic events in the future? No doubt, we can do that, or something like it, sometimes. But Deleuze's claim that some events are not yet determined, but only determin*able*, only works if the future is not determinate, that is, only if the future after what came before is empty. This hardly seems controversial: The future may not be like the past. Deleuze is just pointing out that if such commonsense theses about the future are right, then the future is empty. One might wish the future to include *both* what has not happened yet, *and* where things are actually going. But for Deleuze, that would be a category mistake. The (pure) future is only the former.

Deleuze's phrase "pure and empty form of time" (*DR* 87) says that time has form but no content. In two passages, Deleuze calls time "formless" (*DR* 91, 297), but in context the point is similar. I count nineteen appearances of the phrase "pure and empty form of time." Sometimes it is a predicate: "Death ... corresponds ... to the empty form of time" (*DR* 112); "Thinking ... separates out the pure form [of time]" (*DR* 114); "Delay is the pure form of time" (*DR* 124). Sometimes it is a subject term: "time as a pure and empty form ... undoes the circle [of reminiscence]" (*DR* 91); "the pure and empty form of time ... leads to eternity and the return in time" (*DR* 122).

The text is difficult because Deleuze uses two phrases that could have almost opposite meanings. One is "the empty form of time" (*la forme vide du temps*) (*DR* 88/119). By this phrase, we might think that time is the form of succession, as most people would agree, and that the form of time is simply neutral with respect to what content it has or will have at any given time. Although empty in the sense of being neutral, the form is not necessarily ever empty of all content. Strictly speaking, it is the *form of* time that is empty; actual time need not be empty. But Deleuze's other phrase is "the emptiness of pure time" (*le vide du temps pure*) (*DR* 87/118). By this phrase, it is time that is empty, not just the form that is empty. Time here is not just neutral as to which content fills it. There is a specific form of time, namely the future, that can only be empty (whereas present and past forms of time could be filled, perhaps must be). On this construal, time does not have a single form that is sometimes full and sometimes empty. Empty time has a different form than filled time. What is that form?

It would be tidy if we could say that the future is empty of actual content, but does contain virtual content. After all, the sorts of objects that Eros intends are generally non-actual. Sometimes we do desire things that already exist, although in that case, objects of Eros would not be futural (*DR* 96–7). But when we desire imaginary, symbolic, idealized, not yet invented, or generally, virtual objects, then perhaps in those cases, Eros is the paradigm for future-directed consciousness. And in those cases, perhaps we could say that the future is filled, albeit with objects that do not really exist.

Does the future thus contain virtual objects of desire? Or to put it more strongly, is the future virtual? This idea is appealing, but in Deleuze's usage, it seems a misuse of term "virtual." For the most part, when Deleuze uses the term "virtual" in *Difference and Repetition*, it is in the phrase "the virtual part of the object" (*DR* 209), or "the virtual content of an Idea" (*DR* 206), to refer to the abstract and repeatable part of an object or event as opposed to its factual state at a particular moment in time. In that sense, he says, "the reality of the virtual is structure" (*DR* 209). Deleuze at any rate does not use the phrase "virtual future." He does use the phrase "virtual time" on one occasion later in the book, arguing that structure is genetic: "Every structure has a logical, ideal, or dialectical time" (*DR* 210–11). But in his usage in Chapter 2, a virtual object abstracts a part object from a particular episode (*DR* 100), and places it in a neutral zone of intentionality, where it becomes possible to desire that object even when and where it does not exist. The virtual part of an object is certainly not limited to the object's actual existence at any given moment of time, but that does not necessarily locate the virtual object in the future, and it does not really fill the future with content. Indeed, the temporal neutral zone, where the virtual structures of events drawn from various times coexist, is by Deleuze's definition, the pure past. "Virtual objects belong essentially to the past" (*DR* 101).

What could a purely future object be, anyways? When Deleuze says, "Eros tears virtual objects out of the pure past and gives them to us in order that they may be lived" (*DR* 102), the suggestion is that what starts as a virtual object from the past may be lived in the future in the form of some as yet undetermined object, an "object = X." In this usage, the virtual object is in the past, whereas the object = X is in the future. Interpreting the distinction in this way is fragile, since in one passage, Deleuze refers to them together as the "virtual object (object = X)" (*DR* 105). Nevertheless, the peculiar indeterminacy of the object = X might later offer us a way for time to be both empty and filled with content at the same time.

Deleuze borrows the formula "pure form of time" from Kant, and some good commentaries on *Difference and Repetition* have relied on the Kant-Deleuze connection.[5] But it seems to me that the differences are too great for the comparison to be useful. First, time for Kant is a pure form of experience belonging to a transcendental ego; not so for Deleuze. Second, for Kant, the *a priori* form of time, *prior* to experience, is an empty form, but our empirical experience of time is always filled with content, and in this respect, for Kant, there is no difference between past, present, or future. For Deleuze, time is pure and empty form *only* for the future. Past and present do have events for content, even if they are weirdly distributed, but the future qua future is empty in principle. Third, for Kant, we do not experience the empty form of time as an object in its own right, we only use it to organize the intuitions that we do experience. For Deleuze, we have an actual experience of empty time, namely our experience of the future.

Kant is thus of limited value in interpreting Deleuze, in my view. Similarly, while Deleuze introduces the "emptiness of pure time" by referring to Hölderlin's discussion of the break between past and future in his *Remarks on Sophocles*,[6] this too has only limited use for interpreting Deleuze. Hölderlin is concerned with the poetic caesura: A pause in a line of meter or rhyme.[7] He finds that Sophocles uses caesuras after dramatic revelations to indicate uncertainty as to what will arrive in the future. Hölderlin says that a caesura represents "time and its cut" (Hölderlin 54/55). "At a certain moment, time turns" (64/65). Symbolically, the caesura cuts the God–man couple. Therefore, tragically, man "forgets" himself and God [I am not sure why Hölderlin thinks this follows], and "there remains in fact nothing more than the conditions of time and space" [really?] (64/65). To go farther [too far], Hölderlin says, "God is nothing but time" (64/65). After the break, man does not know what to do afterwards, so "In man, beginning and end no longer harmonize like a rhyme" (64/65) [as if they ever did].[8] To put it briefly, the crisis leaves the tragic characters nothing but an empty future, barely a future at all. Jean Beaufret's "Preface" to Hölderlin's text, which Deleuze also cites, adds that, "man no longer thinks either in backward or forward direction" (18). Beaufret ties Hölderlin's "pure and empty form of time and space" back to Kant's forms of intuition that supply no content (21).

These are all interesting ideas, but Beaufret insists, I think correctly, that one cannot read Hölderlin's remarks about time in Sophocles as a theory of "time in general" (23). To my mind, there is a limit to how far Hölderlin's overwrought account of the caesura helps in understanding what Deleuze means by the future. Still, the point to retain is that when a decision affirms that the future will not be a continuation of the past, it not only constitutes a turning point on the time line, it empties the future of what it was going to be. There is a sense in which affirming something new of the future first empties everything out of the future.

This sounds rather frightening, and right at about this point, Deleuze switches from ontological to psychoanalytic categories. The twenty-page middle section of his forty-three-page discussion of the third synthesis of time (Deleuze *DR* 95–115) discusses Eros and Thanatos. Only a few of these twenty pages refer to the future. This is a difficult section of the text.

To put it simply, this section concerns how one experience—say, a pleasure—can get repeated by another. When pleasures are independent (as they sometimes are in the id), they may not be repeated at all. But when several pleasures have similar or identical objects, or when they have similar effects on a subject, then they are repeated, sometimes even to the point of fixation. To be repeated, pleasures have to be limited by objective considerations, which means they have to get fixed to a reality principle. It is true that by the time we get an object we were hoping for, it will likely not be quite what we expected. Nevertheless, when we aim to reproduce pleasures we have already had, our aim for the future is not empty. Memory keeps the old goals alive, habit keeps them consistent, and intentionality keeps them fixed. This (much condensed) is the first synthesis of time, the expanding present, described in psychoanalytic terms, a mixture of habit and memory: An experience repeats an objectified former present (*DR* 103).

But some experienced objects are not unified, and therefore cannot be consistently identified over time; they remain "part objects." When we repeat the desire for a fragment of an object, we are only recalling an idealized take on the past (*DR* 100). Deleuze calls this a (phallic) symbol of the past (*DR* 103), or a displaced simulacrum of the past (*DR* 105). These objects, and the time segments they seek to reenact, are not accurate reconstructions of actual periods of time, and are not necessarily even identified consciously; they are more like problems than objects, repeatable in different ways independent of the formerly real object, assuming there was one (*DR* 106). Artists, lovers, and psychopaths, in different ways, make this shift to virtual objects (*DR* 107). This is the second synthesis: A series of experiences of virtual objects that coexists alongside the series of experiences of actual objects.

The third synthesis arises once virtual objects lose altogether the feeling of necessity. Contingent, virtual objects are free matters, attaching to different actual objects at different times. The ego is freed to choose whatever objects it wants, to the point of narcissism (*DR* 110). It is at this point in the psychoanalytic version of the three syntheses of time that Deleuze reintroduces "the pure and empty form of time, separated from its content" (*DR* 110), breaking with the past. "We enter into the third synthesis. It is as though time had abandoned all possible mnemic content" (*DR* 111).[9] There is nothing here for Eros to desire. The state where "time is empty and out of joint" is now said to be not Eros, but Thanatos, the death instinct (*DR* 111).

Once time is detached from the backward-looking aspects of habit and memory, it becomes detached from their forward-looking aspects as well. In death, and for those of us still alive, in the death drive, Deleuze says, ego and id "will be [future tense] annihilated" (*DR* 111). The future for us is empty because death is empty. (Deleuze cites Blanchot.) As we do not control the future, the future has no other shape for us than the pure form of time. It has no content, so there is no going forward toward it. Unlike Eros, death disinvests from the past; it is energy "desexualized and without love" (*DR* 111). No virtual objects, therefore no love. No love, therefore no branches on the time line (*DR* 113). Coexisting and/or alternate series no longer resonate (*DR* 111)... living death: a "repetition-eraser." (Deleuze indirectly cites Robbe-Grillet.) Some might call it "eternal return," but nothing comes back (*DR* 115). What is eternal about it is just that it is "interminable" and "formless" (*DR* 115). Even the work of mourning ends. This is the passage where Deleuze says most clearly that, "the ultimate synthesis concerns only the future." The synthesis of the future does not explain what the future is, or predict anything about it. It concerns the future only in that "it announces in the superego the destruction of the Id and the ego, of the past as well as the present" (*DR* 115). The future is a time— that is not questioned in the text. But it is a time scope that erases the content of time.

In the "Conclusion" chapter of *Difference and Repetition*, Deleuze makes explicit the connections between decision, the future, and time out of joint. All three depend on detaching the experience of time from habit and memory.

> If this third time, the future, is the proper place of decision, it is entirely likely that, by virtue of its nature, it eliminates the two intracyclic and extracyclic hypotheses [the cycles of habit and of memory]; that it *undoes* them both and puts time into a straight line, straightening it out[10] and extracting the pure form; in other words, it takes time out of "joint" and, being itself the third repetition, renders the repetition of the other two impossible. (Deleuze *DR* 296)

"This is how the story of time ends: by undoing its too well centered ... circle and forming a straight line which then ... reconstitutes an eternally decentered circle" (*DR* 115). The last part of this spatial metaphor is funny, because circles all do have centers, so time is not really a circle (or a line or a grid). If we have to assign it a geometrical form, we might best say simply that the form of time is the joint ("The hinge is the empty form of time," *DR* 284): filled on one side, empty on the other. What is important anyhow is not the geometry, but the future without memory: seriality without projects.

Death sounds a trifle negative, but there is a positive form of experience made possible by detachment: Thought. "Thought ... opens us on to empty time" (*DR* 114). Indeed, this essentially becomes Deleuze's definition of the third synthesis: Thought, which "abstracts time from all content in order to separate out the pure form" (*DR* 114).

At a certain point in the text, this sounds like Deleuze's conclusion. As formal and groundless thought could be anything, thought is the most futuristic form of consciousness. It proceeds without truth-values, and without time-values. Eros too could be oblivious to past desires, but thought is more so; thought empties itself of everything except its future interpretations, subject to chance and excess. But is the groundlessness of thought sufficient to call the future empty? Does this not repeat the problem: Are groundless thoughts empty? Is Deleuze affirming a kind of decisionism in the realm of thought? Is the future empty if empty thoughts are in it? How is this kind of future "the proper place of decision" (*DR* 296)?

This is a difficult passage to interpret. When Deleuze says here, at the end of his twenty pages on psychoanalysis, but with thirteen pages left to go in the chapter, that "this is how the story of time ends" (*DR* 115), does he mean that this maximally empty time is finally the best definition of time, or is he suggesting (without saying it) that psychoanalysis does not finally generate a good enough theory of time, because the best it can say about the future is that it is a death drive?[11]

... unless the emptiness of the future is a condition for a different kind of content for the future. At this point in the text, psychoanalysis drops out of the picture, and the remaining pages of Chapter 2 of *Difference and Repetition* deal with "dark precursors" and Deleuze's spin on Nietzsche's "eternal return." This certainly sounds like a more productive way to aim at the future than either the death instinct or the indeterminacy of thought. But even with the theory of dark precursors, it is not easy to say what, other than emptiness, the future is about.

We can start by asking: When do dark precursors exist? The point of saying that precursors are dark is to say that even when we believe that an event will mean something in the future, we recognize that we cannot foresee, or decide in advance, the series of repetitions that it will instigate. To be a normal sort of precursor, an event has to have some special property that prescribes something about future events. But if an event is a dark precursor precisely by not clarifying a future series, by pointing to the region of an event that remains undetermined and ungrounded, then in an important sense, dark precursors are the opposite of precursors. It would be nice if we could say that the future is the time when we take up events from the past, and by an act of will, repeat them differently, many times, on different branches, as it were. It would be nice, in short, if we could say that the future is when the "eternal return of the different" takes place. But with dark precursors, any connection between a past event and a future repetition is an "excessive coupling" (*DR* 118–9): Neither the past nor the future elements of a pair explain why they are joined as they are.

The foundational reason why dark precursors do not light up the future is that what is repeated is not an action: "*all* that returns [to the future], the eternal return, is the *unconditioned*," a kind of unformatted problem space (*DR* 297). Past contents can of course be repeated in new presents, which can be repeated in turn in later presents, but "the third repetition renders the repetition of the other two [past and present] impossible" (*DR* 296). No matter how much repetition takes place, and no matter how much is anticipated, the future remains empty. This may not mean that there are no precursors at all, but that dark precursors are dark in a strong sense. A dark precursor is not a precursor of an event that will come later on, but is a dark precursor of itself. It is not a precursor *of* something dark; the precursor itself is dark. It empties *itself* out so that it cannot have deterministic effects. It is a kind of repetition-maker that ensures that its own agency does not yet exist. This is the force of saying that the dark precursor resonates not "in advance," as a normal precursor would, but "in reverse" (*DR* 119), by "retrojection" (*DR* 126). The point is that the singularity of an event, such as it is, the way an event emits its stream of singularities, the differentiator of its differences, exists only in the future. The precursor *is* what has not yet happened.

As it happens, Deleuze's most developed example of a dark precursor is difficult to fit into a theory of the future. The example involves Freud's case of a childhood trauma that lies dormant, and gets actualized in symptoms only later in life.[12] Deleuze calls the childhood trauma the dark precursor (*DR* 124). There are two ways to think about futurity in this scenario. One way is to think that the trauma and its later expression make up together a single event that takes place at two distant times. We might want to call the later stage of the event the future of the earlier stage, but if they are really two poles in the same event, it is more a case of distribution of a single event across time, than of one event and its future. The second stage would be dark at the time of the first, but then again, the first would remain obscure until the second occurs. Because of this, one might equally say that the adult symptom is precursor of the childhood event. In any case, the dark precursor on this construal sounds more like before–after coexistence than like an empty future.

The second way to think about the trauma-to-symptom passage is that the second stage is the "delayed" presence of the first. ("It *is* this delay," *DR* 124). This comes closer to thinking of the second as the future, despite the fact that it is the essence of the first. Whether we know or imagine the second stage in advance does not matter as much as the fact that the event is half future at any given time, that the event depends on the half that is not there yet. The part of the event which is empty would be the sense in which the event is in the future.

The second reading allows us to go back to the question: When is there a precursor of the future? Answer: In the future. To say it simply, the future exists only in the future. Consequently, to perceive the future in its pure state, we would have to exist in our not-yet form. But in a future image, what is there to see?

To sum up, it seems on the one hand natural, if perhaps a dead end, to think of the future as empty, as nothing has happened or is happening in it. It is conceivable that Deleuze's interest in the future simply has nothing at all to do with our everyday chronological future, with what might happen to us later today, or to humans in the coming century. But put this way, it is no so natural after all to think of the future without actualities. Will the price of oil not fluctuate in the future? Will this semester never end? Of course, we do not read Deleuze for his stock market predictions. And it is right for philosophers to distinguish sharply between what it *means* to be future and the particular occurrences that will end up happening in the future. It is tempting to conclude, with Arkady Plotnitsky,[13] that "[Deleuze's] formlessness may be called 'pure time' only insofar as it is no longer 'time' at all in any sense we can give to such a concept, in this respect making the term 'pure' more important than 'time'" (Plotnitsky 140–1). But if possible, it would be preferable to preserve the time in "pure time," the future in the "synthesis of the future."

We want to push the problem of the future beyond the dichotomy of either a temporal void or a classical theory of precursors. In terms of Deleuze's text, the thing to keep in mind is that even though the third synthesis of time involves forgetting past and present to empty out the form of time, nevertheless in time as a whole, the syntheses of the passing present and the coexisting past remain autonomous, and still keep chugging along. Even once we experience the future as empty, the present keeps contracting past and future, as you can plainly see; the past keeps bubbling up into the present and the future, just look; the future keeps emptying out past and present. So in what form do past and present events subsist once the future destroys them? Does simulated content from the other syntheses return, albeit emptily, to the empty future, like the free-floating forking paths in a Borges story?

Deleuze on future tense

Does it mean anything in Deleuzian terms to say an event "will" take place? In the Deleuzian context, it is worth returning to the question of future tense. The linguist Émile Benveniste has the natural view that whereas past tense refers to the actual past, future tense designates only subjective anticipation.[14] But for Deleuze, the future is not just subjective. So what should future tense do? The grammar of natural languages does not decide what the future is, since languages vary in their usage of future tense: Some languages have only a short-term future tense, others treat future tense as an imperative. English does not have a straightforward future tense ending the way French does, it has only an auxiliary "shall + infinitive."

French has both, and Deleuze uses them interchangeably in "Three Novellas," where he says a few things about future tense. By his classification, a novella tells a story about what just happened; a tale is about "what will happen" (Deleuze *ATP* 192–3). Deleuze's interest is in the former, but mine is in the latter. Deleuze's odd example of a tale is Maupassant's "*Une Ruse*," in which a doctor advises a woman by telling her what happened to someone else. It is a stretch for Deleuze to use this as a story about what a person "will" do, and anyways, it is not told in future tense.

But perhaps it is a red herring to expect the future to be expressed in future tense. Despite his great interest in time, Deleuze recommends that when talking about tales about the future: "let us not invoke too strongly the dimensions of time" (*ATP* 193). After all, by analogy, the past need not be expressed in past tense: the Anti-Oedipal phrase "I am all the names of history" (*AO* 21) expresses the multilayered past in present tense. So likewise, when Deleuze speaks of "a new people that does not yet exist" (*WP* 108), this might be a future reference without future tense. However, the analogy is not perfect: The future is irreducible to present tense in a way that the past is not. Pulling history into the present, by saying "I am now all the names of history," enhances the past. But saying "We are already now the people that does not yet exist," while true in a certain manner, weakens the crucial point of the people's nonexistence. Deleuze acknowledges that past and futures tenses are not analogous. When he describes the past as "determin*ed*"—he italicizes the past tense form. When describing the future, he says it is "*determinable*"; not that it "*will be* determined" (which requires a passive mood), nor that it "*will* determine *itself*" (which requires reflexivity). In fact, there is no obvious way that Deleuze could have make his point about the future using future tense. This has to mean something. The indicative future, "the people will come," suggests too much determinacy for the future. Try: The people "might" come. "Let them come." "They will have come." "They are comable." How does one form an authentic future tense? Does future tense destroy future reference? After all, if the future were simply empty, there would be nothing to say about it no matter what the tense.

Reconciling whether future tense sentences undermine the future qua future, is like reconciling the empty future with dark precursors. Given that *Difference and Repetition* Chapter 2 ends with a few pages about simulacra and the power of the false, might we say that the future tense just simulates future propositions; or that propositions about the future are empty, yet simulacra-filled; or that the future is filled, but with simulacra truths? Is there a way to imagine that in Deleuzian branching tense logic, instead of no propositions about the future being true or false, all propositions about the future, even incompatible ones, are true, albeit empty. The traditional challenge to allowing all propositions to be true is that if both conjuncts of a "P & ~P" contradiction are true, then any proposition can be proved. Yet this anathema might be just right for the future, where every possible branch is real, as long as it is not present. The emptier the future is, the more Borgesian branches it can have.

Where game theory considers branches from state to state, perhaps the empty form of time can envision branches from gap to gap, propositions floating freely in meaning space. One implication would be that actualities even in the present and past would already be empty in their insides, just because the future is already extracting their virtual elements and devaluing their facticity—precisely that emptiness would be the synthesis of the future. For Broad, past and present are full of actuality; the future's nonexistence is the temporal anomaly. For Deleuze, the future's nonexistence rubs off on the present and past as well, which get their potential meaning, under-determined as it is, from the indeterminate truth-values of their futures. For Deleuze, the jungle of future branches pre-persists even before events branch. The future removes the autonomy of the actual. It is not so much that as the future becomes present, possibilities undergo branch attrition; it is more that as the present becomes future, actuality undergoes trunk attrition.

The more we pursue this line of interpretation, the more the empty future converges with the idea of the multiple virtual future. This is what we hoped would happen. But now, we need a formula for thinking of the peculiar kind of multiplicity to confer on the futural references of decisional contents. For this, I return to the fragile distinction I touched on earlier, namely between a virtual object and an object = X.

Object = X and decision = X

I am going to use Deleuze's spin on Kant's formula of the object = X to show how a decision diverges into a plurality of virtual time lines.[15]

Deleuze does not use the word "decision" often, as it sounds too similar to "judgment." He thinks of judgement as the work of a voluntary subject who rationally assesses evidence, cuts off deliberation, and imposes an allegedly knowable future on others. But "decision" does not have to mean that; a decision can be involuntary, collective, or rambling. In the essay, "To Have Done with Judgment,"[16] Deleuze says: "A decision is not a judgment ... : it springs vitally from a whirlwind of forces that leads us into combat."

Similarly, in *Difference and Repetition*, Deleuze distinguishes two senses of "decision" (*DR* 115).[17] The bad kind, like judgement, "imposes limits." The good kind expresses "excessive systems which link ... the multiple with the multiple, the fortuitous with the fortuitous ..." (To be precise, we should create a distinction parallel to Deleuze's distinction between differen*c*iation and differen*t*iation—namely between deci*s*ion and deci*t*ion: The decision that actualizes a future, and the decition that problematizes a future.) Decisions are in some ways like "questions" (*DR* 115), because they deal with situations in doubt, but decisions do not aim at an answer in the same way.[18] Decisions are like choices, but need not be voluntarist or teleological. Decisions are like imperatives, but they are more like an imperative to throw dice than to obey.

In short, when Deleuze speaks of the "power of decision" in *Difference and Repetition* (*DR* 197 and 199),[19] he means something like a commitment to a serial problem. A decision is not "decisive" in the sense that an argument or a battle might be. It is not a state of mind, a set of reasons, or an outcome, but a divergence point and all its possible branches, both fulfilled and unfulfilled. Decision is thus neither about the present nor about a particular future: Its peculiar kind of future is in-between present and future. A true in-between is not between two actual times, but has its own kind of time structure. When Deleuze says that, "the future is the proper place of decision" (*DR* 296), he does not mean that having made a decision, we wait to see it carried out. He means that a decision is an encounter with diverging futures. A decision is inside an assemblage of possible time-worlds.

I will call on a complicated history of resources. In Chapter 2, I discussed Husserl's description of decision as a "volitional noema." In *Logic of Sense* (*LS* 20–21), Deleuze uses Husserl's noema to explain "sense." Husserl defines a noema by its open horizon, and following Kant, as an object = X. Kant's = X sounds like a formula for identity, but read through Fichte and Hegel, = X becomes an expanding collector-function. Deleuze describes sense as an object = X, citing Kant and Husserl. He describes decisions as singular points emitting diverging lines into the future. The single decision affirms many possible worlds, like throwing dice. Deleuze calls Adam's decision to sin a "pure game" (*LS* 172): But more like go than dice (*Foucault* 67–8).[20] We will work through these connections.

Turning points

In *Logic of Sense*, Deleuze analyzes decisions as turning-points. Any "object of affirmation" is "prolonged in a line of ordinary points" and then "begins again in another singular" (*LS* 53). Of course, it is difficult to measure how singular a turning-point is, or how dispersed its outcomes are. But this is OK, because Deleuze is interested in decisions as "problems," that is, as networked diagrams of "turning points" (*LS* 55), or "points of inflection" (*LS* 52). Each turning point has its own pattern of interactive elements and implications, and depending on how it functions, might be described in terms of "crisis points, boiling points, knots, and foyers" (*LS* 55). The "field" of such points can be called a "jet of singulars" (*LS* 53); the "field of selection" maps out diverging pathways (*LS* 172).

The central idea is that a turning-point is dispersed into a jet of points. Both the turning-point as a whole, as well as each of its elements, "is defined as the locus of a question" (*LS* 56). The point of decision "communicates" with its possible enactments (*LS* 172). There does not need to be a very first point in a decision chain, or a final point. But the segments of a decision sequence, whether fixed or supple, have to be diagrammable along a "positive distance of different elements" (*LS* 52–3); it has to be possible to follow a series in detail, and to mark provisional starting- and stopping-points for divergences and cut-offs that shift series into different series. A decision is relayed through points, and "*begins again*" at the very next point. Only serially can the next point be a new beginning, rather than a lonely moment (as it is in Sartre). Decisions, and actions generally, have regular consequences as they affect neighboring objects and events, although they diverge into situations that redefine it and the events around it. Futurity in decision-making processes is carried by relays, shaped by direction and vicinity, density and distance, and time to arrival at destinations by different routes.

In a decision's pathway, new decisions interrupt and communicate (*LS* 53). Deleuze puts it slightly differently in his book on Foucault: A series of events is "surrounded by a cloud of ordinaries," and emits a singular point that "can be made to move anywhere" (*Foucault* 60). The "real definition" of an individual describes it as a concentration of pre-individuated possibilities on route to something different (*Foucault* 63). One might describe an individual only by its actual, ordinary continuations, as if nothing were left to be decided, but that would merely be a "nominal definition." Indeed, because there are layers of decisions that are only virtual, and do not appear in the actual series of points before and after the turning point, the seemingly ordinary can often be a radical decision in disguise.

Some of Deleuze's examples of relayed series come from mathematical series, others come from chance encounters, neither of which seem to me ideal paradigms for decision. Still, the theory of turning points can help describe many features of decision-making. To imagine a typical example, when somebody decides to emigrate for a "better life," a decision which may already be confused and disguised by politics, or may intersect with disease control or some other trajectory, this decision has some ordinary continuations—bureaucratic and utilitarian—but may in time generate other decision topoi, intersecting with a decision to change vocation, a turning point that will have its own series, starting with ordinary jobs, then not-so-ordinary divergences around love, justice, or adventure, and so on step by step.

A futurally under-determined decision need not be vague. By way of analogy, Deleuze says that there is nothing vague about assigning an under-determined predicate like "colored" to an object. Deciding to talk about color in general is to take on a visual problem that will be pursued by various investigations, from graphic art to optic nerve repair. Likewise, a decision with under-determined consequences need not be indeterminate, as long as it undertakes to explore futures under a specified predicate. This open continuation is the sort of genesis implicit in what we will call a decision = X.

"Their transformations form a history" (*LS* 53). Or in reverse, each turning point is caused by "a fragment of a future event" (Deleuze cites Péguy, *LS* 53). This is what I have been getting at. The future is not just the outcome of a decision; it is the temporal term already inside the decision content. The future reference, with all the non-actuality that the future implies, is just the empty part of the thought that makes it a decision in the first place.

It may be true that decisions generally go off track, but that fact is anticipated in their sense from the start. Decisions where we cannot know when or how it will get enacted, will not refer to any actual future dates; all of the space-time coordinates of that sort of decisional future are virtual. The driving force of Deleuze's *Logic of Sense*, after all, is to separate "spatio-temporal realization" from an "ideal event" (*LS* 54), to distinguish chronological time from ideal incorporeal time. Turning toward a particular end rarely if ever occurs without also turning toward a larger set of alternatives in the time-multiple, most of which will forever remain nonfactual. Just as irrational numbers constitute the vast majority of number, so irrational time is the vast majority of time. Indeed, because the present is a point on many time lines, most of which will not get actualized, it follows that the present too is already more virtual than actual. In this way, Deleuze can say that the present is "the empty square [although it is empty of content, it is saturated with branches] or the mobile element" (*LS* 55).

The big payoff for us is that a decision has not one but many futures. Deleuze quotes Borges saying, "No decision (*décision*) is final, all diverge into others" (*LS* 61/77). The Borges story that Deleuze cites here is "The Lottery of Babylon."[21] A person wins a lottery; the prize is death, but before he gets the prize, a large but unknown number of further decisions will get made, which might alter the prize entirely. The moral is that the distance from a decision to its effects is increased not only by delay, but more importantly, by subdivisions of the decision into an indeterminately dense number of intervening decisions. This is usually what Deleuze means by the labyrinth of the straight line. When something finally happens, it will be impossible to infer backward to the initial decision that led to it. For a scientific view of decision, it is bad enough if we do not know what future consequences a decision will have; it is worse if we do not know which actions were caused by a given decision in the past, or even if there was any decision in the first place.

Borges' story "The Garden of the Forking Paths," which Deleuze refers to elsewhere, has a more direct conceit that all possible options in any given decision are made and enacted in parallel. The moral of the two stories together is that decisions pile up not just many equally probable branches, but many overlapping actual futures. But Borges writes fiction. Where, if anywhere, does this happen in life? For Deleuze, as we have seen, only one activity in real life can force time to branch into "elongated pasts and futures" (*LS* 62): Thought.

Only thought can devote itself to what will happen in the future, and yet remain "neutral" as to how much time passes before it is achieved. A decision takes off from actual psychic moments, but "becomes autonomous in the act of disinvesting itself from its matter and flees in both directions at once, toward the future and toward the past" (*LS* 62). It comes into existence at the same moment the virtual future becomes its new temporal reference. In fact, the decision is as neutral to the actual past that it re-characterizes, as it is neutral to the actual future. The decision as prior cause, and the future as posterior effect, become interactive, and escape the ordinary run of things, at the same moment.

The reason thought can anticipate many different continuations in a single object, is because a thought object is an object = X. Once we see how there is an object = X, we will see how there can also be a decision = X.

Kant's object = X

The idea of the "transcendental object = X" originates in Kant's *Critique of Pure Reason* (Kant A105–9),[22] in the "A" version of the Transcendental Deduction. Kant names two very different kinds of X: "X (the object)," and the "object = X."[23]

> It is clear that, since we have to deal only with the manifold of our representations, and since that X (the object) which corresponds to them is nothing to us—being, as it is, something that has to be distinct from all our representations—the unity which the object makes necessary can be nothing else than the formal unity of consciousness in the synthesis of the manifold of representations. It is only when we have thus produced synthetic unity in the manifold of intuition that we are in a position to say that we know the object ... This *unity of rule* determines all the manifold, and limits it to conditions which make unity of apperception possible. The concept of this unity is the representation of the object = X, which I think through the predicates. (Kant A105)

The first, "X (the object)," "is nothing to us." This is the thing-in-itself distinct from representations: Simple identity. There is no such thing. The second is the "transcendental object = X," the synthesis of apperception directed toward a singular object. There *is* such a thing, even when the rule of synthesis is "imperfect or obscure" (A106). The synthesis is not itself a representation (we do not experience ourselves undertaking this synthesis), but it produces a "transcendental object = X" (A109). Transcendental apperception (Deleuze calls it the "zone"[24]) makes = X's out of anything. If experience were limited to apperceptions alone, its objects would consist of unordered sensible contents. If experience were unlimited, its objects might include noumena or things-in-themselves. But experience extends to something in-between these two extremes. By synthesizing apperceptions according to rules, its objects consist of complex empirical things. The transcendental act of synthesis makes the sides of a figure, or the shapes and sounds of a thing, cluster around a common target referent. The act of synthesis constructs an intentional object, a convergence point so that we can attribute qualities drawn from various representations to the same referent. And because there is nothing to define the object other than the procedures and materials for constructing it, nothing prevents further properties from being added to the object. In short, synthesis constitutes the empirical object as a divergence point into new properties, and at the same time as a convergence point for previously discovered properties.

Some commentators on Kant focus on other issues to interpret the object = X. For Longuenesse, for example, = X emphasizes rules.[25] For Badiou, = X is the unity prior to synthesis.[26] For Rudisill, it means that the identity of objects is "mentally supplied."[27]

For Heidegger,[28] the object = X represents an epistemological problem with the whole idea of an object. The = X is the gap that arises when we try to know how it is that we know Being, or when we try to make Being into an object. "What is known of this knowing [of beings]? A Nothing. Kant calls it the 'X' and speaks of an 'object'" (Heidegger *Kant* 85). The business of knowledge operates in-between objects and things-in-themselves, in-between what we know and what we do not know. In short, = X refers to the way we do not know the final state of what we know, that we identify objects only in the mode of a "rough sizing up" (*Kant* 87). Heidegger thus thinks of the = X not as a procedure for identifying a singular X, but as a problem with the very idea of a singular object. But the X does not simply name the unknown, it names a task to pursue.

Commentaries that compare Deleuze and Kant tend to focus on issues other than the nature of objects. Simont,[29] for example, compares Deleuze and Kant in terms of schemata; Beaulieu[30] in terms of epoché and world.

As I see it, Deleuze does draw a theory of objects from Kant, and interprets Kant in the best way. For Deleuze, the = X formula names X as a stand-in, so that anything that should arise during the genesis of a synthetic object will get collected and connected without the whole ever being bounded. Deleuze cites four = X functions. (a) = X lets attributes "resonate" (*DR* 117, 122, 124, and 291). (b) = X allows an object to appear in "disguise" (*DR* 105 and 291). (c) = X is motivated by "desire" for synthesis (*DR* 110). (d) = X makes the future present (*DR* 119 and 291–2). Deleuze extrapolates from the perceptual object = X in cognition; to a symbolic object = X in the unconscious; to a portmanteau word = X in language (*LS* 234;

DR 123); and to an action = X in history (*DR* 299). My project extrapolates to a decision = X in ethics.

In his 1978 course at Vincennes,[31] Deleuze interprets Kant to mean that synthesis is the contraction of parts in space and time, grasping simultaneous and successive parts within a single object. "It's from this level of the analysis of the synthesis of perception that Kant can be considered as the founder of phenomenology." The "=" form is obviously not visible in sense-experience, therefore it must be a presupposition prior to sensibility. In one sense, the form of synthesis "refers" to nothing. But it produces an object as an object = X, so it is not simply nothing; it is an "any-object-whatever." A perceiver can only find determinate objects if she first has the object = X form to put that determinacy into, and can produce possible expansions in imagination. She also has to recognize what determination each = X should convert into, namely what predicates to describe it with.[32] = X is never the final description of the object. It exists only to have some determinate content substituted for the X. Yet its form remains an under-determined = X even after it is determined. The object = X thus cannot prevent epistemic catastrophes: We are generally not able to stop adding parts to the object. We might not be able to reproduce the preceding phases of the genesis, or even to recognize the increasing whole. This dizziness is the original sublime.[33]

Kant discusses the object = X only in the A version of the Transcendental Deduction. Perhaps because that version was printed in only a few copies, and most of his contemporaries read the B deduction instead, the object = X did not have much direct effect on philosophers of the day. The post-Kantian philosopher with the most X is Fichte.[34]

Fichte's = X

On the surface, Fichte's A = A is the opposite of Kant's object = X. The former is a dialectic of identity, whereas the latter is a synthesis of a manifold. But identity for Fichte is already a kind of overreaching, an encounter with difference, and this will make Fichte's identity even more synthetic than Kant's. I will review just a few of Fichte's moves.

Fichte's starting point is any act of consciousness whatsoever, the affirmation of some A, and no more than that: A = A. This already takes up two positions toward A: The first posits A; the second posits A as the predicate of itself, as its own self-reflection. A = A is thus something like an object = X, where the object = itself (97–8). Because the second A is not merely the first A, but is to be added, it entails the possibility that the second A will not be the same as the first. In positing the first A, we do not yet know what the second A is, that is, what A's predicates are, so while the second A is epistemologically a requirement of the first, it is also an obstacle to it, an obstacle that takes place within the formula of the object A itself. In other words, the second A is in principle ~ A.

At stake in the generalized = X formula is not just the identity of an object, but how an object can get equated with predicates. For Fichte, what does the equalizing, and so contains the predicates = X under the original object X, is the subject. The object presents as X, but cannot by itself incorporate its = X into itself objectively. Only the subject can do that. On this score, Fichte's subjective idealism is quite opposed to Kant's transcendental empiricism. In this respect, Deleuze sides with Kant.

But Fichte does not stop either with the unity, or the opposition, of subject and object. Fichte's = X is meant to distribute shifting degrees of reality and negation in the object, measuring what the object has in it and what it does not, which elements in the object the subject is free to determine and which not. The vicissitudes of advancing and receding borders between subject and predicate, and between subject and object, drive most of Fichte's work, from the relations of cause and effect to the politics of private property law. The idea that = X is a format for defining and exceeding the limits of objects brings Fichte's = X closer to Hegel's than to Kant's.

Hegel's = X

Hegel (rightly) finds that Kant's simple = X formula is too indefinite, a bad infinite, a mere "also," the bland relativistic assumption that we can add any predicate we like to an object. The ethical parallel would be as if we could carry out a decision by means of any action whatsoever. Admittedly, Kant's = X schemata are not entirely indifferent to what they collect. They combine and anticipate content by quantity, quality, relation, and modality. Still, Kant's = X formula assumes that exactly the same forms of synthesis work for every object, and the formula does not change as divergences occur during the genesis of an object. It does not synthesize the object's identity with its differences, the way Hegel advocates, or express internal differences, or formulate differentials that complicate the object's identity, the way an object = dX/dY (to invent a Deleuzian formula) would.

In the "Perception" chapter of the *Phenomenology of Spirit*, Hegel derives, but then critiques, the attempt to describe an object by naming one predicate, then another predicate "also," and another "also" … It is easy to show that predicates do not simply accumulate by addition, but combine under various forms of tension. Color predicates exclude one another by occupying a surface; color predicates require shape predicates; one object's active predicate (an eagle's claws) correlates with nearby objects' receptive predicates (a rabbit's skin); diachronic predicates are each negated at a later point in time. This is not to dismiss the object = X formula altogether, because an object does hold many predicates. But "=" generalizes a much more concrete genesis and structure of predicate interrelations.

Hegel's twist in his *Science of Logic* on "The Thing and its Properties" (Hegel *ScL* 484–98) challenges the separation between Kant's two senses of X. Kant distinguished the propertyless thing-in-itself from the synthetic empirical object defined by properties. Hegel argues that as the empirical object is flexible enough to contain many properties, it divorces itself from any particular set of properties, presents itself as an empty substrate, and so the empirical thing too becomes thing-in-itself-like. And as for the property, since it is free to be instantiated in a plurality of things, it is as independent of particulars as the thing is, and so becomes a property-in-itself. The thing-in-itself status proliferates to the point where every phenomenal difference is a micro-thing-in-itself. Of course, the upshot for Hegel is not to confirm the thing-in-itself, but to show the absurd consequences of the ontology of existing things and their properties.

The ultimate consequence of the fact that things are neutral with respect to which properties they can absorb, and that properties are neutral with respect to which things have them, is that things can have the properties of *other* things as easily as they can have their own. ("It is through property that the thing is continuous with other things," *ScL* 493.) It is not just that a stem and a leaf can both be green, but that the greenness of the stem is transferred to the leaf, then later to the finger that rubs it, and so on. Causality is the distribution of properties from one thing to another. Hegel concludes that instead of properties, we should speak of "free matters," like pigments and electrical forces (and, we might add, *petit à* desires, and ideational memes). "Matters circulate freely out of or into 'this' thing" (Hegel *ScL* 494.) This gives a very different model for the object = X. The object here is as much a disseminator of, as a collector of, properties. The formula =X is a field for the circulation of properties across space, time, and substances. This logic brings Hegel into view of Deleuze. For Hegel, = X represents a turning point, where individuals turn universal, and things turn into other things.

Hegel gives two main arguments to show that identity implies difference, that is, that if $A = A$, then $A =$ some other X also. First, identifying A requires isolating it from what is already a manifold (*ScL* 414–5). Therefore A has relational predicates prior to self-identity. Second, while the *linguistic* proposition $A = A$ seems to say nothing new in the predicate, the *thought* proposition $A = A$ has a cognitive sequence, namely A ... [wait for it] $= A$. The thought "sets out to say something more", and is disappointed (it "contradicts itself") when it ends up saying nothing else (*ScL* 415). Therefore, again, there is a place-holder for predications different than A in the form $A = A$. As a static predication, it is true that identity takes the essentialist form of $A = A$; but as a genetic predication, identity takes the synthetic form, which we have called object = X.

That a predicate other than A is found in the object = A, requires immanent development, which is just what Hegel calls the "concept." A predicate judgement, A is B, implies the conditional judgement, If this is A then it is B (*ScL* 654). This is turn implies: If an object achieves all the properties involved in A, then it gains all the properties involved in B. Therefore, the more we know about how A is B, the more we know a series of conditionals, which articulate the gradual unfolding of A into B. Several such hypotheticals are needed to define A by its various disjunctive possibilities. Only as the object (or our knowledge of it) is actualized, does its disjunctive definition take the form of a conjunctive judgement. Activated hypotheticals compile the results achieved so far, in the object = X form: A is B and C. To the extent that the object's determination is not complete (and its completeness, Hegel says, is never more than contingent), the final judgment regarding the object remains disjunctive. The object, in short, is equated with a self-expanding and self-limiting (inclusively and exclusively disjunctive) set of predications.

In the final analysis, then, predication is found not in the vicissitudes of the S is P form, but in the object = X form, expressing a history of difference in rigid and supple objects. The more we see predicative judgments as expanding and contracting spheres of influence over time, the more the object = X form converges with a decision = X form.

Deleuze's = X

I have lingered over the logic of = X so that as we return to Deleuze's account of decision as a turning point whose future has an = X relation, we have a subtle logic of "=" to work with, reducing neither to identity nor to additive conjunction, but capturing and releasing flows and interruptions of predicate series over time.

The first step in connecting =X to the topic of decision is to apply the object = X model to acts of will (a step Fichte takes in his *Science of Rights*).[35] We need to redescribe = X in two ways that Kant did not: As a future time = X, and as an act of will = X.

Deleuze comes close to calling the future an = X (Deleuze *DR* 103), emphasizing that = X remains empty even once actualized. A plan that can be fulfilled in ways other than intended, is an "action = X" (*DR* 110).[36] Since an action can always have further development, a future time = X is more, later, than any particular moment in future time. Yet at the same time, since the endless future is always cut at a finite and dated point when some action is taken, the future time = X is also less, earlier, than future time. There is thus a double meaning when Deleuze says that "the object = X is the immanent limit of the series of virtuals" (*DR* 109). The = X belongs to the series of virtuals, and it is a limit-point of virtuality. In our terms, a decision = X can either defer action, or condense it in a hurry. To say that a decision is not an empirical enunciation but an object = X is to say that it does not so much individuate a decision content and aim at a future reference, as deconstruct a meaning and set several jets of points streaming forward.

In this context, Deleuze says that the object = X is a "floating signifier and floated signified" (*LS* 228). What "happens" on one series "insists" on another (*LS* 228). The X is "displaced" because it diverges into different possibilities, and spreads out over virtual time, because after all, its meaning is its genesis. A decision is open both on the time lines of possible fulfillment as well as on the actual time line. A decision in this way is not just an image existing in the present representing something meant to happen in time. But a decision is nevertheless an image of time, an idea about what time means, about what the future is, and about when the future is not. In his essay "*À quoi reconnaît-on le structuralisme?*," Deleuze describes this double articulation of being both outside chronological time and yet introducing a temporally unfixed event into chronological time. "[The object = X] is perfectly determinable, even in its displacements ... It is just not fixable in an identifiable place" (*À Quoi* 264). By its genetic structure, an object = X permits a person or a society to find "a certain number of possible choices," and to find, "at each of their points of choice [or at each moment in history], a certain number of possible individuals" (*À Quoi* 265). It is not a stretch to regard an object = X, defined as a genesis of choices, as a decision = X.

We need to say more about this double feature of being in virtual and chronological time in different ways. Insofar as it has no straightforward identity of temporal reference, we could call the = X function "pre-sense" (since it precedes actualization); or we could call it "nonsense" (since it includes mutually exclusive branches); or "co-sense" (since it describes neither a present not a future state but a duration that includes both, not to mention also the times in between, before, and after) (*co-sens*, *LS* 233). The point is that when contemplating a decision, "we temporarily abandon life, in order to then temporally fix our gaze upon it," on a different time scale not centered on our present preferences (*Nietzsche* 25).[37] After all, in decision, we are not just choosing one set of results among others, but get involved in a range of conditionals built on further conditions. So what is the temporal meaning, the temporal noema, of a decision?

Husserl's = X

Deleuze introduces his own concept of sense through "Husserl's reflections on the 'perceptual noema'": Sense is "the ideational objective unity as the correlate of the act of perception" (Deleuze *LS*, 20–21). Husserl in turn introduced noema by way of Kant's object = X. Whereas a *noesis* is the subjective description of an experience, the object meant is the "determinable X in the *noematic* sense" (Husserl *Ideas 1* 313/270).[38] The object will only have been partially determined at any given time, but it is always further determinable. The noema must remain an "empty X" (*Ideas 1* 315/272), in order to be open to new perceptions: "The pure X in abstraction from all predicates" (*Ideas 1* 313/271). This is important, since for Deleuze, determinability is what grounds time in Kant (*DR* 86).

Husserl and Deleuze have different ways of spinning this point, but the important elements are shared. Husserl says: "in no noema ... can the pure determinable X be missing" (*Ideas 1* 315/272). Deleuze says the determinable X *is* missing (Deleuze *LS* 228), in the sense that X is under-determined and under-identified. But even for Deleuze, the missing X is always what the experience is about, and in that sense, the missing X is what is not missing. To say it simply, for both Husserl and Deleuze, an object is an =X, but the full meaning of it is missing. Following Aristotle's view that the mind must be neutral in order to be formable by any object whatever, Husserl says that the empty X is "the possibility of harmonious combination to make sense-unities of any level whatever" (Husserl *Ideas 1* 315/272). If we leave out the constraint of "harmony," Deleuze can agree. A "plurality of rays" of meaning "go to the X with the highest synthetic unity" (*Ideas 1* 316/273). In *Analyses Concerning Passive and Active Synthesis*,[39] Husserl says the "substrate X" describes how a horizon of an object's determinability follows an "unceasingly changing sense." Only if the sense of an object is open to change is it sure to be able to describe the whole object, however the object might show itself. This is an important thesis about openness and determinacy. Commonsense assumes that closure means determinacy and openness means indeterminacy. But closing off the sense of an object is always premature, and guarantees that the sense will leave aspects of the object unstated. Contrary to commonsense, a closed definition always leaves an object indeterminate, whereas specifying ways in which the object is always still open is what gives determinacy over time to its objective sense (*Passive Synthesis* 58).

Sense is in this sense future tensed, or genetic. Deleuze complains that Husserl's noema ought to, but cannot, describe both a meaning, and its future genesis. Deleuze says—I think wrongly—that the kinds of predicates that Husserl thinks define meaning cannot prefigure which subsequent events will confirm them (Deleuze *LS* 97). For a meaning to engage as yet unknown objects, the noema must be a hybrid of meaning and effect (*LS* 98–9), which Deleuze thinks reduces Husserl's framework to nonsense. For Deleuze, = X articulates just this necessary nonsense. In fact, I think that Husserl does have a place for a meaning that projects its own future options, even to the point of frustration and deformation. He calls it a noematic "modification," expressing an intentional "perhaps," a modal meaning: "'a volition-meaning,'" a "noema belonging peculiarly to the willing" (Husserl *Ideas 1* 233/199), precisely a decision noema. It seems like Deleuze ought to agree to the "perhaps"-intention, and the decisional noema.

But first, we need to take a detour through a different, and in my view less successful, line that Deleuze sometimes takes on decisional futures, in which he describes decision less like noematic genesis, and more like throwing dice.

Dice throw

Deleuze suggests the dice throw as a paradigm action = X. He says the pure gamer throws without preferring any outcome. It is as if an object is so open to determinations that affirming the object simply means taking a chance and accepting whatever turns up.

On this account, for a game-player, or decision-maker, genuinely open to the future, the goal is not to test a strategy, or to test a description of an object, holding out for the definitive or correct result. Only a "bad player counts on many throws of the dice," waiting for probabilistic patterns to impose themselves on chance (Deleuze *Nietzsche* 26–7). The game is to play just once. It is true, of course, that if you gamble your car on a single dice throw, you will not get a second chance. But Deleuze's criticism of bad players goes farther: Those who gamble on a particular answer or number are playing an impure game—they are not really playing a game at all, they are working a formula (*LS* 58–9). An impure game exhibits (a) preexisting rules, (b) statistical probabilities, (c) if ... then options evaluated by skill and effort, (d) distinct sequences of plays, and (e) gains or losses. Game Theory deals with such decisions, and we could think of chess, or corporate gamesmanship, as paradigms.

There is, after all, some plausibility to the Game Theoretical idea that games are only interesting when skill outweighs luck, and chance is controlled. After all, a game of pure chance, like craps, is the most boring activity possible—just throw, that's it—it only gets interesting if the dice are loaded. Nevertheless, Deleuze thinks that "pure games" are not based on calculated risk (*LS* 59–60). Pure games: (a) invent rules, (b) ramify the effects of chance, (c) play out in simultaneous constellations, (d) are nonsensical, and (e) have no winners or losers. It seems in these passages that for Deleuze, the only games that avoid statistics-based decisions are games where there are no decisions at all.[40]

Deleuze offers several paradigms of the pure game, and sometimes it is hard to see what they have in common. The pure game of dice emphasizes chance. The pure game of Thought emphasizes freedom to invent rules. Sometimes he names Go as a pure game, which emphasizes flexibility (although as players know, flexibility is achieved by hard choices, not by chance). But it might not be that certain games are essentially computational and others essentially the opposite. It may be that any game can be played as an example of each. Marriage arrangement is usually a measured risk, but it is not impossible to arrange a marriage as free play; while Go, or even Thought, are generally nomadic, but can be played by the computer. If we had to choose the purest game under the criteria given earlier, thought is certainly a better example than dice. Yet because Deleuze uses the dice throw motif in several texts, it is worth saying something about it.

In my view, the dice throw model of decision overplays chance and underplays time.[41] In particular, the dice throw model underplays the contribution of seriality to freedom. Deleuze says that a dice throw affirms chance "every time" (*DR* 198), and that "the whole of chance is ... there in a single time." There are two issues here. Is "chance" the right concept for describing the virtuality of a decision's relation to the future? And is the decision's whole relation to the future marked "in a single time?"

Let us assume that outcomes of a decision are subject to chance and cannot be controlled. What does that mean for the temporal focus of that decision? It could mean (a) that the decision can only care about the moment of decision that it controls, and must be indifferent to whatever chance outcome arises; or (b) that the decision affirms the possible outcomes, and treats the moment of decision as trivial. The pleasure of the game might be (a) in throwing the dice into the air, or (b) in seeing an unpredicted number turn up on the table. In my view, the decision has to cover both ends, even if one end or the other or both cannot be controlled. A decision without predictable outcomes means a commitment to getting-there no matter where; a decision without effective agency means a commitment to getting-there no matter how. The dice throw is an odd paradigm, but insofar as the two directions of uncontrollability together put past, present, and future all in play, the pure game is temporally consistent.

When Deleuze says, "The future is the proper place of decision" (DR 296), the minimum it could mean is that decisions separate from the past, from prototypes and obligations. For this point, the dice throw is a proper analogy, since each throw is independent of past throws. But this much is reminiscent of decisionism. The downside of the dice throw paradigm of decision-making is that it has no effect on the next throw. A throw has virtually no effect; worse, it has no virtual effect. After all, we cannot throw, in "a single time," anything like a philosophy, a child, or any interesting noema.[42] There has to be a better way to get free of past actualities than just to throw the dice, if a singular point is going to emit a jet of particulars.

And why insist that a thought-throw does not take place over the course of many throws? What is interesting about a throw of thought, unlike a throw of dice, is that it emits loose ends. Of course, when a person starts a thought project and waits to see where it goes, it could be described as a single chance-taking act with many possible ends, just as one indivisible project has many futures, or one indivisible life contains many events, or one individual subject contains many experiences. But these are classical one–many relations. Why insist that the dice throw, or destiny, is decided just once? There is no good reason to believe that purity of heart means to will one thing. Whether throwing dice is like accepting one's fate, or like creating the world by playing at drafts, or like subjectivating oneself, why insist on a single throw?

Of course, singularity for Deleuze means more than individuality. Singularity is complicated, because it is distributed through a field into destination objects, subdividing and reassembling (*LS* 113–4). "Each thought forms a series in a time which is smaller than the minimum of consciously thinkable continuous time. Each thought emits a distribution of singularities" (*Nietzsche* 60). We might say that even a single throw is a multiple all at once: "unity affirmed of multiplicity," "multiple affirmation" (*Nietzsche* 29). In fact, Deleuze uses two different terms in *Logic of Sense*: There is one "toss" (*lancer*) of the dice, but there are many "plays" (*coups*) (*LS* 113/138).[43] This allows us to think of a single decision split up into many stages, including options for restarting the process or revising past selections. But it still regards the decision to make decisions as a single originary toss-up, and this sounds more like Heidegger than like what Deleuze ought to say. Why should the thrower, who is after all more than thrownness, whether gambler or schizo-nomad, be content with one roll per assemblage?

One way to sort out the one-and-many of decisions is to say that there is just one dice throw-like decision, but that it affords many interpretations. "The interpretation of the eternal return begins with the dice throw but it has only just begun. We must still interpret the dice throw itself at the same time as it returns" (*Nietzsche* 31). But interpretation sounds like it adds a second decision, another dice throw.

Another approach is to say that a single decision collects a plurality of forces. "The will to power is never separable from particular determined forces, from their quantities, qualities, and directions" (*Nietzsche* 50). Chance and force might seem opposites, but "by affirming chance, we affirm the relation of all forces" (*Nietzsche* 48). The emphasis is no longer on the flight of the dice, but on the vectors in the hand that throws. In *Cinema 2*, Deleuze cites the casino scene in the abysmal Jerry Lewis movie (Godard and Truffaut loved it[44]), *Hollywood or Bust*: Jerry's body "is shaken by spasms and various currents ... when he is going to throw the dice" (*Cinema 2* 65). The chance effect is now located earlier than the throw, in the preexisting forces that mess it up. Jerry wins a lot of money, but his jerky movements are out of all proportion with moving the dice around. The throw is propagated by a wave of energy, rather than by any one particular pathway. The thesis that there is only one throw, one chance, one decision, or one becoming, is more palatable if the forces collected in a "single" decision are already plural beforehand. The trajectories of the forces and vibrations within the decision may be too complicated for the eye to make out. But under this picture, micro-temporal sub-movements are the structures for the decision.

In short, the decisional throw may appear in one shot, but is in truth made of a "diversity of coexisting cycles" (*Nietzsche* 49). When Deleuze says of dice that "the number of the combination produce repetition" (*Nietzsche* 26), this seems unlikely to mean that we repeat one decision. He must mean "repetition for itself" in the sense of *Difference and Repetition*: multileveled temporality. This trajectory returns us to what I think is Deleuze's better model of decision than the dice throw, the futural = X model. How can we connect the singularity of a decision, made up of coexisting temporal cycles and waves, with a decision = X?

In *The Fold*,[45] Deleuze's case study for the decisional future extrapolates from Leibniz's analysis of Adam's decision to sin. Coexisting temporal cycles will now be articulated as coexisting possible worlds.

Adam's decision and incompossible world cycles

Deleuze never quite says that Adam chooses to sin, as Leibniz does. But Deleuze and Leibniz agree that when Adam sins, he branches into one of many possible worlds. Possible worlds provide a semantics for an object = X, and have in fact been so used to interpret Husserl's noema.[46]

On Deleuze's telling, all versions of Adam start in the Garden, then diverge when Adam makes his decision (*LS* 114). When an object = X like Adam splits, "another world begins" (*LS* 111). Since it is not the individual, but its versions, that begin worlds, it follows that "all possible situations occur" (*LS* 114). That is, if there is something that some version of Adam could do, there exists a possible world in which he does it. Some highly similar versions of Adam may be equally possible in the same world. But an Adam who sins is incompossible with a world in which Adam does not sin. An Adam who did not sin, for example, would have led to a world where people are extremely good, and that is definitely not our world. Yet this irreducible difference between options is just what makes Adam's decision into a real divergence point at the moment it occurs in the real world. Deleuze is not saying that incompossible worlds all exist in the same *actual* world. That is still impossible. And it is not that incompossible Adams live their whole lives in different worlds, since that would leave each Adam no choice—each Adam in a given world would only be able to be the Adam consistent with that world, he would have no freedom to do otherwise. Different Adams can neither live entirely in the same world, not entirely in different worlds. Different Adams have to start off in the same world, and then diverge into different worlds.

Although incompossible worlds thus cannot exist in the same *actual* world, freedom requires that the paths diverging from decision coexist in the same *virtual* world. In Deleuzian modal realism, all worlds virtually occur.

David Lewis and possible worlds

We can compare Deleuze's thesis about the reality of incompossible worlds with David Lewis's[47] modal realism, which also holds that all possible worlds are real.

When Lewis says that only one possible world is actual, all he means is that only one world is indexed to a given person's life. Every possible world is equally a world. But each person lives in one world only, and that is the world that is actual for him. Nobody lives in more than one possible world, either simultaneously or at different times. Lewis says, "I cannot think of a single philosopher who favors trans-world identity," that is, who favors the idea that a given person "leads double lives" (198). But Deleuze does favor it. For Deleuze, the same Adam is projected, from a certain time onwards, into different worlds. For Lewis, the most we can say is that very similar Adam "counterparts" live in different worlds. In many of those worlds, the Adam counterparts are almost exact duplicates up until Apple Time, but then become very different from their counterparts after Apple Time (assuming we can match up Apple Times across worlds, as "temporal cross sections ... counterpart centuries, or weeks, or seconds," Lewis 71). But for Lewis, when the Adam counterparts diverge, it is not one Adam splitting. It is only the several counterpart Adams, who have been living in different worlds all along, becoming more discernable.

For Lewis, all that happens is that independent worlds become increasingly different. He calls this "divergence," not branching (206). "Branching," Lewis thinks, is a crazy theory. For Lewis, it is absurd to think that the same Adam branches into different Adams, that the same world branches into different worlds, that the same time line branches into different time lines, that one present has more than one future. He thinks it absurd that following a decision, different futures branch out of the same person. Lewis thinks he can refute the theory of branching by pointing out that "it conflicts with our ordinary presupposition that we have a single future" (207). What planet does Lewis come from? I, for one, do not presuppose that I have a single future. Deleuze, in any case, believes in what Lewis calls branching, although Deleuze calls it divergence.

Although Lewis would reject the ontological and temporal multiplicity at the heart of Deleuze's philosophy, he does open the door to one very Deleuzian problem: In a universe made of many possible worlds, how does an individual know in which world, at what time, and even which person, she is? The problem is that individuals can only know which world they are in by knowing truths about their world (27–30). But since the truth a person knows might hold for many similar worlds, different possible worlds are "accessible" to a person in a given world. Therefore, a person cannot know which world she is in. Furthermore, each person has a counterpart in those other worlds, so she cannot know which, of many individuals, is her. If, as Deleuze thinks but Lewis does not, decisions branch, then at the moment of decision, there is a sense in which we are pre-individual. And because for related reasons, we cannot know how far along we have gone on a decision's time line (a problem that Decision Theory acknowledges), this makes us, we could say, pre-chronic.

The incompossible worlds that Deleuze speaks of result from diverging futures that become incompatible with each other. Strictly speaking, incompossibility is not a relation between two versions of one person, or between two worlds, but between a person and his other possible world. It describes the relation between Adam the sinner and the world in which Adam the non-sinner exists (Deleuze *Fold* 59). Incompossibility occurs—no surprise—when "series diverge in the neighborhood of singularities" (*Fold* 60).[48]

Deleuze sometimes describes entry into incompossible worlds as a "divine game." His paradigm game this time is not dice, but a "calculus of infinite series ruled by convergences and divergences" (*Fold* 61). In *Logic of Sense*, Deleuze suggests yet a different paradigm game, when he says that "the ideal player of the game is Aion," as if time itself is the player (*LS* 64), something like the way Heraclitus says that time is a child playing at draughts. And in the same passage, he suggests still another paradigm game: "[The game] plays ... at the border of two tables" (*LS* 64), with an "impenetrable window" connecting them. This sounds like Duchamps' "The Bride Laid Bare by her Bachelors, Even," played as a board game.

Whatever the image, Deleuze's idea is that the first move of a game decides the rules of the new world; as it were, on an empty board. In *The Fold*, Deleuze names Go as if it were Adam's game. (Leibniz knew something about Go,[49] but whether he understood all the rules is unclear.) Of course, Go has rules, but where chess begins with situated and determinate pieces, Go begins with an empty grid. The winner of a game of Go is the player who has taken the most territory (the most empty space), but paradoxically, you cannot get enough territory if your strategy is to take territory. Go strategy involves building groups whose essential meanings are designed to be able to shift. Takemiya's "moyo" strategy, for example, is to claim an impossibly large territory, sacrifice it as soon as his opponent invades, surround the invader at a distance, redirect power elsewhere, converting that power to a new territory only at the last minute. Deleuze says that in Go, "you encircle your adversary's presence to neutralize him, to make him incompossible, to impose divergence upon him" (*Foucault* 68; *ATP* 352–53). I think it might be better to say you try to make your opponent's stones inflexible, merely compossible, deprived of divergence.

In fact, Go has a technique specifically suited for incompossible world analysis, that is, for situations where an individual is incompossible with a world not his own. In effect, it is a strategy for making decisions in games one is not playing in. Usually, the result of a game sequence is judged good if each move leading to it was optimal, that is, if we chose each divergence with the best outcome. But in "Tewari" analysis, we imagine a different sequence that could have led to the same final situation on the board.[50] If not every move in that unplayed series was optimal, then neither was the played series optimal, even though *its* every move was. Criteria are thus not inside a series, but inside its counterparts. Good plays are hidden in those other possible game worlds. Even to say that Go players play through thousands of hypothetical games in their mind at once, only one of which shows up on the board, is not the half of it. The power of a decision is measured not just by the other worlds where the same decision led to different results, but also by the other worlds where different decisions led to the same result. The moral of the incompossibility game is that within a given series, decisions extract and activate elements from alternate temporal cycles, which are running contemporaneously, but discontinuously, in the various possible worlds they initiate.

Now, coexisting hypothetical series make for interesting game playing, and interesting decisional noemata; but a bad move in a game is no sin, original or otherwise. The motive to play has no ethical value, so games do not ultimately suit Leibniz's concern with Adam's sin, which, Deleuze says, is to provide "the first great phenomenology of motives" (*Fold* 69). In fact, Leibniz says little of Adam's motives, and in the event itself, the Biblical Adam shows little imagination for the future consequences of his choice. The Scripture's entire account of Adam's decision reads, "She gave her husband some and he ate it" (*Genesis* 3:6).[51]

Nevertheless, Deleuze's view is that decisions are motivated, indeed motivated "in *every* direction" (*Fold* 69). His odd proof is that motives are simply images of possible futures (*Fold* 70). To put it bluntly, we have images of many incompossible futures, therefore we are motivated to make many incompossible choices, therefore we make many incompossible choices, therefore we bring about an incompossible object = X. The Go player has in fact played out all the games she has imagined playing out; a decision-maker diverges into all the possible futures that the decision's noema puts into superposition.

Deleuze at this point compares Adam's decision with one of his own: "For example, I hesitate between staying home and working, or going out to a nightclub" (*Fold* 70). We do not usually picture Deleuze at a nightclub, but he pictures it. As the images of "the hum of the word processor," or the drink in the bar, respectively, vary in intensity, inclination swings back and forth. (This sounds like

William James, see Chapter 1.) Sometimes the motive is determined merely by "where my region goes the furthest"; that is, sometimes we incline to whichever image projects the longest future (*Fold* 73). On this theory, motives will vary over time, as moment by moment, the balance of perceptions might incline the same agent in different directions. To persist without substitution, a decision would have to be "renewed" at the next moment (*Fold* 71). Lots of temporal structures are possible for a decision, and elements of each may or may not be rational. But the point is that a motive is just a futural noema. Referring to the future in a certain way is just what turns a neutral image into an active decision.

At each moment, then, a decision image projects all the perceptions one has. This is what we mean by saying that a decision is autonomous. Autonomy means completing the job of one's own decision, leaving none of one's stones unturned. Yet the moment is the divergence point, so deciding in the moment, the subject enters all his possible futures at once. The best autonomy is heterography.

Controversial ontological theses like this need arguments, so here is one: If we lived only one of our decisional options, then we would never be present at the moment of divergence. We would always be on the path we were on, and never be at a point which lies on two different paths, either of which we could from that moment proceed on. Indeed, there would never be a point where divergence could occur. Lewis would be right, and we would never make a decision. In contrast, if the present *contains* the divergence into two paths, the decision-maker at the instant of decision is already on both paths. This does not prove that we make decisions; but it proves that if we make one decision, we make them all. This is the being of the becoming: The synthesis of time *in which* the present passes, the present as it exists on many levels. And once we are on both paths, how could we believe we follow one all the way and the others only a short way? We may live some futures with lower intensity, or with less awareness, than others (the way all monads reflect the same totality of events, but in different ways and to different degrees), but we live them.

What did Deleuze decide? We know he finished writing his book on Leibniz. We know he took the occasional drink. We do not know which he did at that moment, but he did both over time. Can each of us say that while we made certain decisions, we made the other ones too, that there need be no anxiety or despair since we live all the lives we cut off? There are several ways to say this: We make all possible decisions; or, the paths we decided against, we lived out anyhow; or, decisions made and decisions not made are superimposed on different but coexisting time lines; or, we actually make one decision but the others still affect us virtually; or, we never actually decide, since decisions exist in virtual time only.

Obviously there are challenges to such a theory. For one thing, even if thought makes multiple decisions, bodies seem to act just one way at a time. To be sure, bodies too have virtual and ambiguous modes. Many corporeal decisions—fight or flight, harm or cure, laugh or cry, sink or swim, eat or burp—follow their own kinds of branching patterns into coexisting though incompossible worlds. Still, bodies generally do not display all their possibilities at once the way thought does. If a body only lasts a certain time, to take the clearest case, then at least for some purposes, decisions about it only have so much future. Decisions resulting in somebody's death, for example, clearly limit variation (although the finality of death is not heavily marked in Deleuzian chronology). We need to concede something along these lines. After all, we should not expect the temporality of decisions about thought and decisions about body to have exactly the same forms. But this does not necessarily undermine the theory of multiple decisional futures, it just reinforces the thesis that every kind of decision has its own futural structure. The difference between decisional noema and decisional behavior is itself a source of temporal difference inside poly-linear decision-making in the syntheses of the future. All structural differences of these sorts overlap in the intentionality of futurity presupposed by the attitude of decision-making. There are as many temporal structures as we can decide to make; they are under-determined, they are determinable, they are collected in the decision = X, and the future is filled by all of them.

Afterword

The future is the concern of decision-making, and yet it is not actual. For some philosophers of time, the future has to be as determinate as the past and present, but for a person making an open-ended decision under uncertainty, it is not. Some philosophers say that the future is not actual, but possible, but for a decision-maker, the future is more than a mere possibility. At the very least, it is what Hegel and others call a "real possibility," or a material or causal possibility, rather than a "formal possibility" (Hegel *ScL* 546–50). To the extent that we hold ourselves responsible for what happens, we treat the future not just as a logical possibility but as the world where we and others are really going to live. When we make a decision to do something, we have to see at least some pathways leading from where we are to what the future is intended to become. But decision must be more than causal possibility. A decisional noema may create possibilities that were not causally possible before, and it may well exclude most of the available causal possibilities. We have to empty out the time line there was, and enter non-real decision-objects into the temporal gap opened up. But we need to really do that. The intended future has to be realistic, no matter how stretched and different from the present. The future of a decision is neither an empty hope, nor a random fluctuation, nor a deterministic causal projection. It is both more and less than possible.

For a similar reason, a decisional future is not just a potentiality grounded in the actual present. It is true that a plurality of potential futures could be grounded in a single present. But defining a potential depends on being able to determine the state of the actual present, and we have found that the actual present, understood as a turning point, is in part defined by the futures it can become. To know what is potential in a present requires knowing which stage, or node, of which branching decision one is at in the present. To decide by way of the potential future what to write next, one should know whether one is currently at the end of writing the second book of a trilogy before the start of the third; or if one is about to retire and start a new career; or something else. A decision-maker can be at all sorts of stages in all sorts of series. This does not harm the decision-making practice or its futural noema. But it does mean that there is no descriptive or explanatory value in regarding decisional futures as potentialities grounded in the present. Instead, the futures intended in a decision have a force of their own to describe the whole series of decisions and subdecisions between (and including) the present and any chosen future point. It is in this sense that the pluralization of future branches and series transfer the future from the ontology of actuality, possibility, and potentiality, to the ontology of virtuality. The virtual does not have to refer to anything especially bizarre, unpresentable, or surreal, nor does it have to arrive by chance or by surprise. To be virtual, the future just has to be plural at every step.

The back cover blurb for science fiction writer William Gibson's *Distrust that Particular Flavor* reads: "Instead of predicting the future, [Gibson] finds the future all around him, mashed up with the past."[1] I agree that the future is not set off from the present by a unique order, and I agree (Althusser and others say this too) that every present is anachronistic, mashed up with elements of past and future. But I do not think that the whole of the future is already present (Gibson does not assume that either). The future, on the account I am aiming at, would be polypresent: becoming present in fits and starts, in series with branching and recurring nodes.

To flip this conclusion around, it means in conclusion that decisions are directed at actions only in the virtual future. And if events in the virtual future are themselves virtual events, this means that we can only decide to perform virtual actions. Of course, once we do them, the actions are actual. But what we decide to carry out is the virtual act. And to go a step further, if decision is the mode of intentional act in which we make ourselves ethical agents, and if decision is the primary act for which we are responsible (since, at least relatively speaking, we have more control over how we decide than we have over the effects of our behavior), it follows that what we are responsible for are events in the virtual world. In short, ethics pertains to the virtual. Or to put it again even more strongly, we are ethical only to the extent that we switch our attention from the actual world to the virtual.

It might seem ethically problematic to restrict decision to virtual time: at worst, such decisions commit to no particular behavior, but to endless time line fantasies. Such decisions might seem both too casual (as they fail to settle down) and too burdensome (as they call us to act in many possible worlds at once) to count as ethical responsibility. But virtual decision-making stands a better ethical chance than decision's two other extremes: namely, Sartre's view that decisions have no future, and Decision Theory's view that decisions have nothing but. For Sartre, a decision to stop gambling has no future, as the will cannot limit future will; the volitional noema cannot endure new situations. For Decision Theory, in reverse, decision normally calculates by backward induction from a future value. But desiring the best future leads, in the Prisoner's Dilemma, to betrayal, which does not give the best future. Players do better when they do not know how many subdecisions remain before the outcome will be settled, and so have time to negotiate; decisions are more rational when future branches are under-determined. The multiple futures of a decisional noema are pathways along which decisional objects, namely futures, trade off predications.

Philosophy up until now has only tried to understand the future. Is the point to change the future? Could it mean anything to set out either to change the events in the future (no events exist in the future at all, so how could we change them?), or to change what it means for there to be a future, to empty out the present more thoroughly than usual? When Deleuze says that it is thought that abstracts the future, it sounds like the same thing to think the future, and to change time so as to make it more future.

I have decided to write my next book on the relation between the immediate future and the proximate present. The book will be titled, *Short Term*.

Sun Ra: "The best thing is to consider time as officially ended. We work on the other side of time."[2]

Sekiyama Riichi (First Honinbo Go Champion): "Play every move as if it were the first in your life."[3]

NOTES

Introduction: What is the temporal reference of a decision?

1. Slavoj Zizek. *For They Know Not What They Do: Enjoyment as a Political Factor*. London: Verso, 1991: 222.
2. G. E. M. Anscombe. *Intentions*. Ithaca: Cornell University Press, 1957.
3. Andy Hamilton. "Intention and the Authority of Avowals." *Philosophical Explorations* 11, 1 (March 2008): 23–37.

1 Sartre: Decisions and the unbound future

1. Jean-Paul Sartre. *Notebooks from a Phony War 1939–40* (translated by Quintin Hoare). London, UK: Verso, [1940] 1999. My page references are normally to the translation, but when I insert my own translations, I refer to page numbers both to the translation and to *Les carnets de la drôle de guerre*. Paris: Gallimard, 1983. Unless otherwise noted, all references to Sartre are to this text.
2. Michel Sicard touches on wartime food in the *War Diaries*, but does not discuss the diet. "*La nourriture du philosophe*", *Études Sartriennes*, no. 15, 2011: 61–90, especially 76.
3. Jean-Paul Sartre. *Being and Nothingness* (translated by Hazel E. Barnes). New York: Philosophical Library, 1956: 31–37. Hereafter *BN*.
4. Robert Sokolowski. *Eucharistic Presence: A Study in the Theology of Disclosure*. Washington DC: The Catholic University of America Press, 1994.
5. "Options ... are the noematic correlates of projects." The "noema ... motivates perceptions."
6. Karl Jaspers. *Philosophy* V. 2 (translated by E. B. Ashton). Chicago: University of Chicago Press, 1970: 158–160.

7   Thomas C. Schelling. *The Strategy of Conflict*. New York: Oxford University Press, 1963: 22.
8   Simone de Beauvoir. *The Ethics of Ambiguity* (translated by Bernard Frechtman). New York: Citadel Press, 1991: 26–27.
9   This implication follows when we place together these two claims: "Human reality is moral because it wishes to be its own foundation"; and "Man is a being who flees from himself into the future."
10 Jennifer Bates discusses the problem, in Kierkegaard and in *Hamlet*, that oath-taking winds up in a vicious circle of always needing to be taken again. "Hamlet and Kierkegaard on Outwitting Recollection," in *Shakespeare and Continental Philosophy* (edited by Jennifer Ann Bates and Richard Wilson). Edinburgh: Edinburgh University Press, 2014: 40–55, especially 46–48.
11 Friedrich Nietzsche. *On the Genealogy of Morals* (translated by Walter Kaufman and R. J. Hollingdale). New York: Vintage Books, 1969.
12 "[Freedom] aims to suppress itself by nihilating the Nothingness it contains. The ideal of freedom is thus ... a possible that would immediately be an *excuse*. The intimate dream of all freedom is suppression of the hiatus between motives and act ... From this moment on, 'it's not my fault.' Thus every excuse invokes Necessity" (Sartre 134–5/170).
13 Vladimir Jankélévitch. *Henri Bergson*. Paris: Presses Universitaires de France, 1989: 50.
14 William James. *The Principles of Psychology*, V. 2, chapter on "Will." Mineola: Dover Publications, [1890]1950: 560.
15 William James. "Draft on Brain Procedures and Feelings," in *Manuscript Essays and Notes*. Cambridge: Harvard University Press, [1872] 1988: 247–255.
16 William James. "The Will to Believe," in *The Will to Believe and Other Essays in Popular Philosophy*. Mineola: Dover, 1956: 1–31.
17 Benoît Denis has a note on delay in the *War Diaries*. "*Retards de Sartre*", *Études Sartriennes*, 2005, no. 10: 189–209, especially 193–194.
18 The same desire for mastery that empowers the decision, weakens it. Jean-Pierre Boulé, one of the few commentators to mention this passage, interprets Sartre's attempt at mastering his body as an attempt to fill the existential void at an uncertain moment of the war. *Sartre, Self-Formation and Masculinities*. New York: Berghahn Books, 2005: 97.
19 Silke Panse. "The Judging Spectator in the Image," in *A Critique of Judgment in Film and Television* (edited by Silke Panse and Dennis Rothermel). New York: Palgrave Macmillan, 2014: 33–70.

20 See, for example, Felix Ó Murchadha. "Love's Conditions: Passion and the Practice of Philosophy," in *Thinking About Love: Essays in Contemporary Continental Philosophy* (edited by Diane Enns and Antonio Calcagno). University Park: Penn State University Press, 2017: 81–97, especially 90.
21 Theodor W. Adorno. *Negative Dialectics* (translated by E. B. Ashton). New York: Continuum, 1973.
22 Jean-François Lyotard and Lean-Loup Thébaud. *Just Gaming* (translated by Wlad Godzich). Minneapolis: University of Minnesota Press, 1985.
23 John S. Mbiti. *African Religions and Philosophy*. New York: Anchor Books, 1970.
24 The degree of synthesis among time lines can vary. In 1938, one might have thought there were independent options for eliminating Hitler: either renewing German culture, or instigating proletarian struggle. By 1940, it seemed necessary to do all such things "*simultaneously*" (Sartre 296).
25 For analysis of this problem in Aristotle and Derrida, see my *Simultaneity and Delay*, London: Continuum Books, 2012: 150–152.

2 Husserl: Decisions and temporal overlap

1 Edmund Husserl. *Ideas Pertaining to a Pure Phenomenology and to a Phenomenological Philosophy, First Book* (translated by Fred Kersten). Dordrecht: Kluwer Academic Publishers, 1982. *Ideen zu einer reinen Phänomenologie und phänomenologischen Philosophie*. Tübingen: Max Niemeyer Verlag, 1993.
2 Edmund Husserl, *Experience and Judgment* (edited by Ludwig Landgrebe, translated by James S. Churchill and Karl Ameriks). Evanston: Northwestern University Press, 1973. *Erfahrung und Urteil*. Hamburg: Felix Meiner Verlag, 2014. When I insert translations of my own, I cite both texts. Unless otherwise noted, references to Husserl in this chapter are to this text.
3 Alain Berthoz. *Reason and Emotion: The Cognitive Neuroscience of Decision Making* (translated by Giselle Weisse). Oxford: Oxford University Press, 2006.
4 Michel Henry. *Phénoménologie matérielle*. Paris: Presses Universitaires de France, 1990.
5 John Drummond gives a similar account of the noema of future times. John J. Drummond. *Husserlian Intentionality and Non-Foundational Realism: Noema and Object*. Dordrecht: Kluwer Academic Publishers, 1990: 167 and 245–248.

6   Meinong (1910) introduces the term "quasi-object" to deal with a related issue, which he calls "the decision-question". In asking a narrowly defined question about whether a certain objective statement is true or false, the questioner does various things. He expresses a desire to know; he manifests a "knowledge-feeling"; he awaits a response with the intention of making his own judgement once the response is forthcoming; he makes assumptions about the situational context in which the question makes sense and is worth asking; and in some but not all cases, his articulation of the question indicates that he is leaning toward one of the possible answers. But one thing he does not do is intend an object situation. Meinong notes a subtle distinction between the simple decision-question "Do you think the object-situation X is true or false?", and asking "Consider the object-situation X ... what do you think?" The latter posits an object and asks about it. But the former is not committed to an object situation, and is poised somewhere between the affirmation and the denial that it is a real situation at all. Because a decision-question takes this third position on the object, Meinong suggests we refer to the content of such a question as a "quasi-object" (93). Alexius Meinong. *On Assumptions* (translated by James Heanue). Berkeley: University of California Press, 1983: 89–93.
7   G. W. F. Hegel. *The Difference Between Fichte's and Schelling's Systems of Philosophy* (translated by H. S. Harris and Walter Cerf). New York: SUNY Press, 1998.
8   See John Bruin. *Homo Interrogans: Questioning and the Intentional Structure of Cognition*. Ottawa: University of Ottawa Press, 2001.
9   William Ockham. *Predestination, God's Foreknowledge, and Future Contingents* (translated by Marilyn McCord Adams and Norman Kretzmann). New York: Appleton-Century-Crofts, 1969.
10 Ockham is thinking of Scotus' *Opus Oxoniensis* I, d. 39, Appendix A. In their translators' notes to Ockham's text, Marilyn McCord Adams and Norman Kretzmann argue that this is not a likely interpretation, 72-3n.

3 Heidegger: The original decision to decide

1   Martin Heidegger, *Contributions to Philosophy (From Enowning)* (translated by Parvis Emad and Kenneth Maly). Bloomington: Indiana University Press, 1999. *Beiträge zur Philosophie (vom Ereignis)*. *Gesamtausgabe*, Band 65. Frankfurt am Main: Vittorio Klostermann, 1989. References to Heidegger in this

chapter are generally to this text and translation, unless otherwise noted. Sometimes I cite both English and German texts.
2 Martin Heidegger, *Introduction to Metaphysics* (translated by Gregory Fried and Richard Polt). New Haven: Yale University Press, 2000.
3 The Emad and Maly translation reads "Why Must Decisions Be Made?" (Heidegger 71). I would be happy to find that Heidegger at least occasionally says that decisions are "made". But the German reads "*Warum müssen Entscheidungen fallen?*" (Heidegger 103). I do not have a good suggestion for translating this, but it does not sound like decisions are made.
4 James Scott Baho's *Heidegger and Deleuze: The Groundwork of Evental Ontology* (PhD Dissertation, Duquesne University, 2016), one of the best treatments of decision and time in the *Beiträge*, demonstrates additional elements of this ambiguity. On the one hand, Heidegger thinks of decision as beginning with "hesitant withholding" (242), which implies something of a sequential future. And Heidegger does say that a decision is captured by "limits" (246), and divided into temporal segments. On the other hand, Baho shows that "the operation of the event by which the future becomes temporal and spatial" is not its determinate promise, but the general fact that the event is "differentiated from itself" (238). The event's "abyss of difference exceeds the logic of temporality and spatiality" (241). The "temporal loop" of expectation and memory is "not a duration, but a movement of the genetic 'temporality' of the event" (245). No matter how segmented a decision's expectations are, it produces "temporality," not temporality.
5 Martin Heidegger. *Being and Time* (translated by John Macquarrie and Edward Robinson). New York: Harper and Row Publishers, [1927] 1962.

4 Kierkegaard: Decisionism in religion. Infinite futures

1 See Theopedia.com, entry on "Decisionism." I thank Adam Langridge, a doctoral research assistant in Philosophy at the University of Guelph, for pointing me to this and related sources.
2 Søren Kierkegaard. *Concluding Unscientific Postscript*, V. 1 (translated by Howard V. Hong and Edna H. Hong). Princeton: Princeton University Press, 1992.
3 Søren Kierkegaard. *Philosophical Fragments* (translated by Howard V. Hong and Edna H. Hong). Princeton: Princeton University Press, 1985: 9–22.

4 Blaise Pascal. *Pensées* (translated by H. F. Stewart). New York: Modern Library, 1967: 117–9.
5 Deleuze's game of dice puts an ontological difference between dicethrow and dicefall, making the throw interesting and the result trivial; Pascal's game looks for the correct result, but puts an infinite amount of time between them, making the difference between dicethrow and dicefall longer, but of lesser effect. Gilles Deleuze. *The Fold* (translated by Tom Conley). Minneapolis: University of Minnesota Press, 1993.
6 Immanuel Kant. *Critique of Practical Reason* (translated by Lewis White Beck). Indianapolis: Bobbs-Merrill, 1956: 126–28.
7 From "The Song from the Tomb of King Intef". Also: "Follow the feast day, forget worry!" from "The Dispute Between a Man and his Ba". Both in *Ancient Egyptian Literature: A Book of Readings. Volume 1: The Old and Middle Kingdoms* (edited and translated by Mirian Lichtheim). Berkeley: University of California Press, 1975: 175 and 196.
8 Robert Sokolowski. *Moral Action: A Phenomenological Study.* Bloomington: Indiana University Press, 1985.
9 Don Lodzinski. "The Eternal Act." *Religious Studies* 34, 3 (Spring 1998): 325–352.
10 Luis Molina. *On Divine Foreknowledge: Part IV of the Concordia* (translated by Alfred Freddosa). Ithaca: Cornell University Press, 1988: q. 14, a.13, Disputation 52 and 53: 337–338.
11 Jean-Luc Marion. *Being Given: Toward a Phenomenology of Givenness* (translated by Jeffrey L. Kosky). Stanford: Stanford University Press, 2002. "Immanent Decision," 304–308.
12 Jean-Louis Chrétien. *The Unforgettable and the Unhoped For* (translated by Jeffrey Bloechl). New York: Fordham University Press, 2002: 125.

5 Schmitt: Decisionism in politics. Sovereign moments. Habermas: Steering procedures and the term limits of a decision

1 Matthew Burch. "Death and Deliberation: Overcoming the Decisionism Critique of Heidegger's Practical Philosophy." *Inquiry* 53, 3 (June 2010): 211–234. Burch's defense of Heidegger against the charge of decisionism hangs on interpreting Heidegger's account

of "resoluteness" not as "self-choice" or bootstrap voluntarism, but as following deliberative standards. To use one of Burch's examples, I should decide to make dinner tonight because my wife made dinner last night (213). This example does not sound much like Heidegger.

2   Andreas Kalyvas. "From the Act to the Decision: Hannah Arendt and the Question of Decisionism." *Political Theory* 32, 3 (June 2004): 320–346.
3   Colin Wright. "Event or Exception?: Disentangling Badiou from Schmitt, or, Towards a Politics of the Void." *Theory and Event* 11,2 (2008), online. Although there are many similarities between Schmitt's exception and Badiou's decision, the latter undermines the authority of the leader.
4   Carl Schmitt. *Political Theology: Four Chapters on the Concept of Sovereignty* (translated by George Schwab). Chicago: University of Chicago Press, [1934] 2005.
5   Giorgio Agamben. *Homo Sacer: Sovereign Power and Bare Life* (translated by Daniel Heller-Roazen). Palo Alto: Stanford University Press, 1998.
6   Judith Butler. *Precarious Life: The Powers of Mourning and Violence*. London: Verso, 2006.
7   Walter Benjamin. "Critique of Violence," in *Reflections* (edited by Peter Demetz, translated by Edmund Jephcott). New York: Schocken, 1986: 277–300.
8   Étienne Balibar. "*Sur les concepts fondamentaux du matérialisme historique*," in *Lire le Capital* (edited by Louis Althusser et al.). Paris: Presses Universitaires de France, 1965: 419–568.
9   J. L. Austin. *How to do Things with Words*. Cambridge: Harvard University Press, 1975.
10 "Pataphysics will examine the laws governing exceptions, and will explain the universe supplementary to this one." Alfred Jarry. *Exploits and Opinions of Dr. Faustroll, Pataphysician* (translated by Simon Watson Taylor). Boston: Exact Change, [1898] 1996, 21–22.
11 Benjamin N. Cardozo. *The Nature of the Judicial Process*. Lexington: CreateSpace Independent Publishing Platform, [1921] 2012.
12 Theodor W. Adorno. *Negative Dialectics* (translated by E. B. Ashton). New York: Continuum, [1966] 1973: 32.
13 Jacques Derrida. *Politics of Friendship* (translated by George Collins). London: Verso, [1994] 1997.
14 Jürgen Habermas. *Legitimation Crisis* (translated by Thomas McCarthy). Boston: Beacon Press, 1975.
15 Jürgen Habermas. *Between Facts and Norms* (translated by William Rehg). Cambridge: MIT Press, 1998.

16 Niklas Luhmann. *Risk: A Sociological Theory* (translated by Rhodes Barrett). New York: Aldine de Gruyter, 1993.
17 Jon-Arild Johannesen, Johan Olaisen, and Bjørn Olsen. "The Philosophy of Science, Planning, and Decision Theories." *Built Environment* 24,2/3 (1998): 155–168.
18 "For every one ceases to inquire how he is to act when he has brought the moving principle back to himself and to the ruling part of himself." Aristotle, *Nichomachean Ethics* (translated by W. D. Ross), in *The Basic Works of Aristotle* (edited by Richard McKeon). New York: Random House, 1941: 1113a11, page 971.
19 Niccolò Machiavelli. *The Discourses* (translated by Leslie J. Walker). Middlesex: Penguin Books, 1970. Book 3, section 9, 430–432.
20 David Hume. *David Hume's Political Essays* (edited by Charles W. Hendel). Indianapolis: The Bobbs-Merrill, 1953.
21 Jean-Jacques Rousseau. *On the Social Contract* (translated by Donald A. Cress). Indianapolis: Hackett Classics, 1988.
22 Steve Paikin. *Public Triumph Private Tragedy: The Double Life of John P. Robarts*. Toronto: Viking Canada, 2005, 125. I thank Jamie Smith, Instructor of Philosophy at the University of Guelph-Humber, for this reference.
23 Juan J. Linz. "Democracy's Time Constraints." *International Political Science Review* 19,1 (1998): 19–37.
24 Philippe C. Schmitter and Javier Santiso. "Three Temporal Dimensions to the Consolidation of Democracy." *International Political Science Review* 19,1 (1998): 69–92.
25 Dennis F. Thompson. "Election Time: Normative Implications of Temporal Properties of the Electoral Process in the United States." *American Political Science Review* 98,1 (2004): 51–64.
26 Peter Szendy. *Kant in the Land of Extraterrestrials: Cosmopolitical Philosofictions* (translated by Will Bishop). New York: Fordham University Press, 2013.
27 Katie Paterson. *Future Library*. www.futurelibrary.no.
28 Nick Foton. "Introduction," in *Contingent Future Persons: On the Ethics of Deciding Who Will Live, or Not, in the Future* (edited by Nick Foton and Jan C. Heller). Dordrecht: Kluwer Academic Publishers, 1997: 1–8.
29 For a nice summary of arguments, see Joseph R. Des Jardins. *Environmental Ethics*. Belmont: Wadsworth, 2012: 74–92.
30 Lukas H. Meyer. "More Than They Have a Right to: Future People and Our Future Oriented Projects," in *Contingent Future Persons: On the Ethics of Deciding Who Will Live, or Not, in the Future* (edited by Nick Foton and Jan C. Heller). Dordrecht: Kluwer Academic Publishers, 1997: 137–56.

31 John Rawls. *A Theory of Justice*. Cambridge: Harvard University Press, 1971. Chapter 5, section 45, "Time Preference", 293–298.

6 Decision theory: Seriality effects on decision

1 I draw many definitions from Martin Peterson. *An Introduction to Decision Theory*. Cambridge: Cambridge University Press, 2009.
2 I also draw many definitions from Shaun P. Hargreaves Heap and Yanis Varoufakis. *Game Theory: A Critical Introduction*. London: Routledge, 1995.
3 Michael McDermott. "Are Plans Necessary?" *Philosophical Studies* 138, 2 March 2008: 225–232.
4 See Jay Lampert. *Simultaneity and Delay*. London: Bloomsbury Publishers, 2012.
5 Philip Gerrans. "Mental time travel, somatic markers and 'myopia for the future.'" *Synthese* 159 (2007): 459–474.
6 Thomas C. Schelling. *The Strategy of Conflict*. New York: Oxford University Press, 1963, 173–175.
7 In a Centipede Game, there are two pots of money, which increase in value with each move. Suppose that at the first move, one pot contains $1 and the other contains $10. Player A has the first move. He has two choices: he can choose to end the game right then and distribute the pots, thereby taking $10 for himself and giving $1 to player B; or he can choose to pass, which allows Player B to make the second move. Both pots then increase fivefold, so player B now has the option of ending the game and distributing the pots, so she gets $50 and A gets $5; or passing. Assume that the game can last for as long as six moves. By the sixth move, the two pots are worth $3,125 and $31,250 respectively. It will be B's move, so B will get the $31,250. For A, the $3,125 he would get on move 6 is far more than he would get on move 1. But if A stops the game on move 5, rather than passing, he will get $6,250; this is more than the $3,125 he would get on B's move 6. Therefore A will not let the game get to move 6; he will stop the game at move 5. But B knows all that, since B and A share the "Common Knowledge of Rationality". B sees that she will get $625 on A's choice at move 5, whereas she would get $1,250 on her own choice at move 4. So B will stop the game at move 4. A knows that … so A will stop the game at move 3. Therefore B will stop the game at move 2, therefore A will stop the game on move 1! Even though A will get far less money than if he

passes on move 1 and B wins the game at move 6, A's rational self-interest will result in his receiving $10 upon ending the game on move 1. Every player of a Centipede game will end the game on the first move, due to backward induction.

8   Bartol, Jordan and Stefan Linquist. "How do Somatic Markers Feature in Decision Making?" *Emotion Review* 7,1 (2015): 81–89.
9   H. Rommelfanger. "Fuzzy Decision Theory" (no year listed). www.wiwi.uni-frankfurt.de/profs/rommelfanger/index/dokumente/Aufsatz%20EOLSS%206.5%202002.pdf.
10 Claude Henry. "Investment Decisions Under Uncertainty: The 'Irreversibility Effect.'" *The American Economic Review* 64, 6 (1974): 1006–1012.
11 George Wu. "Anxiety and Decision Making with Delayed Resolution of Anxiety." *Theory and Decision* 46, 2 (1999): 159–199.
12 The Center for Technology and Behavioral Health, Dartmouth College. http://www.c4tbh.org/the-center/what-we-re-up-to/active-projects/flexible-decision-making-in-response-to-unplanned-events (accessed 2014).
13 Eric W. Stein. "Improvisation as Model for Real-Time Decision Making," in *Supporting Real Time Decision-Making: The Role of Context in Decision Support on the Move* (edited by Frada Burstein, Patrick Brézillon, and Arkady Zaslavsky). New York: Springer, 2010: 13–33.
14 Paul F. Berliner. *Thinking in Jazz: The Infinite Art of Improvisation.* Chicago: University of Chicago Press, 1994.
15 Dirk Bunzel. "The Rhythm of the Organization: Simultaneity, Identity, and Discipline in an Australian Coastal Hotel," in *Making Time: Time and Management in Modern Organizations* (edited by Richard Whipp, Barbara Adam, and Ida Sabeus). Oxford: Oxford University Press, 2002: 168–181.
16 For this point, Bunzel (176) quotes E. Esposito. "The Hypertrophy of Simultaneity in Telematic Communication." *Thesis Eleven* 51 (1998): 17–36.
17 Jay W. Forrester. *Urban Dynamics.* Cambridge: MIT. Press, 1969.
18 Jay W. Forrester. "Counterintuitive Behavior of Social Systems." *Technology Review* 73, 3 (1971): 52–68. Citation from *web.mit.edu* (accessed August 2016, 26).
19 For a brief history of the evidence against equilibria, see Adam Curtis's documentary, *All Watched Over by Machines of Loving Grace* [Film]. BBC Productions, 2011.
20 Daniel Peris. *The Strategic Dividend Investor.* New York: McGraw Hill, 2011. I owe several of the references in this section to conversations with Daniel Peris.

21 Daniel Kahneman. *Thinking, Fast and Slow*. New York: Farrar, Straus and Giroux, 2011.
22 Nassim Nicholas Taleb. *The Black Swan: The Impact of the Highly Improbable*. New York: Random House, 2007.
23 John Maynard Keynes. *The General Theory of Employment, Interest and Money*. Cambridge: Cambridge University Press, 1936. Reprinted by BN Publishing, 2008. Chapter 12, "The State of Long-Term Expectation," 97–107.
24 The business reporter Justin Fox gives a good summary of these problems. *The Myth of the Rational Market: A History of Risk, Reward, and Delusion on Wall Street*. New York: Harper Collins, 2009.
25 Philip E. Tetlock and Dan Gardner. *Superforcasters: The Art and Science of Prediction*. New York: Crown Publishers, 2015.
26 Michael Redmond, live commentaries (five hours each) of the five games between the Google computer program AlphaGo and Lee Sedol 9-Dan, March 9–15, 2016. https://www.youtube.com/watch?v=vFr3K2DORc8 (accessed March 15 2016).
27 See the Wikipedia article, "Go and Mathematics." https://en.wikipedia.org/wiki/Go_and_mathematics (accessed January 23, 2017).
28 Alain Badiou. *Number and Numbers* (translated by Robin Mackey). Cambridge: Polity Press, 2008: 107–108.
29 Joe Haldeman. *The Forever War*. New York: Ballantine Books, 1974.
30 The history of go has gone through at least three revolutions up to now: when go was brought from China to Japan in the seventh century and it was no longer required that the first four moves be played on the 4-4 star points; when Dosaku emphasized the center board during the Edo period (seventeenth century) in Japan; and when Go Seigen radicalized central influence in the 1940s. Redmond's view is that AlphaGo is introducing a fourth revolution in go strategy, and that new ideas will emerge in unforeseeable directions.
31 Li Zhe. "Lee Sedol's Strategy and AlphaGo's Weakness", and "Nobody Could Have Done a Better Job than Lee Sedol" (translated by Y. Tong, Chun Sun, and Michael Chen). massgoblog.wordpress.com (posted March 11 2016; accessed March 20 2016).
32 David Silver et al. (Google DeepMind). "Mastering the Game of Go with Deep Neural Networks and Tree Search." *Nature* 529 (28 January 2016): 484–489.
33 Wikipedia. "Monte Carlo tree search." https://en.wikipedia.org/wiki/Monte_Carlo_tree_search (accessed April 12 2017).

7 Branching futures: Tense logic and multiple worlds

1. Huw Price. *Time's Arrow and Archimedes' Point: New Directions for the Physics of Time.* Oxford: Oxford University Press, 1997.
2. Leibniz, Gottfried Wilhelm. "On the Ultimate Origination of Things" (translated by David Garber and Roger Ariew), in *Discourse on Metaphysics and Other Essays.* Indianapolis: Hackett Publishing, [1697] 1991: 41–48.
3. David Lewis, *On the Plurality of Worlds.* Malden: Blackwell Publishing, 1986.
4. Many advocates of temporal symmetry in the philosophy of physics appeal to special relativity's discovery that there is no universal simultaneity across the universe. We all know this as a fact: simultaneity is calculated from a particular frame of reference. Two events might appear simultaneous from one frame of reference and successive from another. The conclusion drawn by the Block Universe theory of time is that there is no sense to the idea of the "present," or the "now," since X might see an event as his past or future while Y is seeing it as his present. The problem with this argument is hidden in what could be meant by "while." The presumption is that X is seeing something as past at the same time as, or simultaneous with, Y seeing it as present. But this is the very sense of simultaneity that relativity rejects. Y does not see anything simultaneous with X seeing something. X cannot see Y's future as X's present "while" it is still in Y's future. So the Block Universe theorist is right to say that there is no simultaneity across the universe when all observers are seeing the universe; but wrong to conclude that there is one *totum simul* at which all events take place. The whole point of relativity is that there is no such thing as time outside of the many frames of reference from which time-order is calculated. Indeed, each frame of reference is defined precisely in terms of what is present, past, and future within that frame, that is, in terms of which events are or are not simultaneous with which others within that frame. The proper conclusion to draw from special relativity is that every conception of time is asymmetrical with respect to present, past, and future. Although there are transformation equations that calculate what observers in other frames see, time for every single frame of reference is inherently about the passing present, and this is the only position from which there is time at all. There have been many construals of Bergson's take on special relativity, but this to me is its upshot. *Duration*

and Simultaneity (translated by Mark Lewis and Robin Durie). Manchester: Clinamen Press, 1999.
5   Paul Horwich, *Asymmetries in Time: Problems in the Philosophy of Science.* Cambridge: MIT Press, 1987.
6   Aristotle. "*De Interpretatione,*" in *The Basic Works of Aristotle* (edited by Richard McKeon, translated by E. M. Edghill). New York: Random House, 1941.
7   Nuel Belknap and Michael Perloff. "Seeing to it that: A Canonical Form for Agentives." *Theoria* 54 (1988): 175–199, especially 190–193.
8   Richmond H. Thomason. "Indeterminist time and truth-value gaps." *Theoria* 36 (1970): 264–281.
9   Rachael Briggs and Graeme Forbes, "The Real Truth about the Unreal Future," in *Oxford Studies in Metaphysics, V 7* (edited by Karen Bennett and Dean W. Zimmerman). Oxford: Oxford University Press, 2012: 257–304. Page references here are to the online version at http://www.marcsandersfoundation.org/wpcontent/uploads/paper_Briggs_and_Forbes.pdf
10 C. D. Broad. *Scientific Thought.* London: Kegan Paul, Trench, Trubner, 1923, Chapter 2, "The General Problem of Time and Change," 53–84.
11 Henri Bergson. *Matter and Memory* (translated by N. M. Paul and W. S. Palmer). New York: Zone Books, 1991: 162.
12 Storrs McCall. *A Model of the Universe: Space-Time, Probability, and Decision.* Oxford: Oxford University Press, 1994.
13 Michael Dummett. "The Reality of the Past," in *Truth and Other Enigmas.* Cambridge: Harvard University Press, 1978: 358–374.
14 See Vladimir Jankélévitch. *Henri Bergson.* Paris: Presses Universitaires de France, 1989: 20–22.
15 Gilles Deleuze and Felix Guattari. *Anti-Oedipus* (translated by Robert Hurley, Mark Seem, and Helen R. Lane). Minneapolis: University of Minnesota Press, 1983.
16 David Braddon-Mitchell and Caroline West. "Temporal Phase Pluralism." *Philosophy and Phenomenological Research* 62,1 (January 2001): 59–83.
17 Norbert Wiener. *The Human Use of Human Beings: Cybernetics and Society.* New York: Doubleday, 1954.
18 Maureen Turim, *Flashbacks in Film: Memory and History.* NY: Routledge, 1989, p. 217.
19 Gregory Currie. *Image and Mind: Film, Philosophy, and Cognitive Science.* Cambridge: Cambridge University Press, 1995. Chapter 7, "Travels in narrative time", 198–222, especially "7.1 Tense in film", 198–206.

20 John McTaggart Ellis McTaggart. *The Nature of Existence*, V. 2. Cambridge: Cambridge University Press, [1927] 1968. See my *Simultaneity and Delay*, 106–112.
21 Pier Paolo Pasolini. *Heretical Empiricism* (translated by Ben Lawton and Louise K. Barnett). Bloomington: Indiana University Press, 1988.
22 John Chu. "Thirty Seconds from Now," in *The Time Traveler's Almanac* (edited by Ann and Jeff VanderMeer). New York: Tor, 2014: 215–221. Originally published in *Boston Review*, 2011.
23 Charles Simak. *Time is the Simplest Thing*. New York: Crest Books, 1961.
24 Cited in Paul J. Nahin. *Time Machines: Time Travel in Physics, Metaphysics, and Science Fiction* (second edition). New York: Springer-Verlag, 1999: 275.
25 John Brunner, *Timescoop*. NY: Dell, 1969.
26 Paul J. Nahin. *Time Machines: Time Travel in Physics, Metaphysics, and Science Fiction* (second edition). New York: Springer-Verlag, 1999.
27 J. J. C. Smart. "Is Time Travel Possible?." *The Journal of Philosophy* 60, 9 (1963): 237–241.
28 David Deutsch. *The Fabric of Reality: The Science of Parallel Universes—and its Implications*. New York: Penguin Press, 1997. Also, David Deutsch, "Interview, in *The Ghost in the Atom: A Discussion of the Mysteries of Quantum Physics* (edited by P. C. W. Davies and J. R. Brown). Cambridge: Cambridge University Press, 1986: 83–105.
29 For an excellent philosophical explanation, see Arthur J. Cunningham. "Branches in the Everett Interpretation," in *Studies in the History and Philosophy of Science Part B: Studies in the History and Philosophy of Modern Physics*. 46, Part B, (May 2014): 247–262.
30 Lisa Randall. *Unraveling the Mysteries of the Universe's Hidden Dimensions*. New York: Harper Perennial, 2006.
31 Murray Gell-Mann. *The Quark and the Jaguar: Adventures in the Simple and the Complex*. New York: W. H. Freeman, 1994: 145–150.

8 Hegel: Morality without decision. Derrida: Indecision Theory

1 G. W. F. Hegel. *Phenomenology of Spirit* (translated by A. V. Miller). Oxford: Oxford University Press, 1977. Section number references are to Miller's translation, but some of the translations

are my own, based on G. W. F. Hegel. *Phänomenologie des Geistes*. Frankfurt: Suhrkamp, 1973.
2 There are two ways to count the three selves. (a) One-duty self; many-duty self; the self who ignores its many-duty self. (b) Legal self; self-recognized by others; conscience self (Hegel section 633).
3 G. W. F. Hegel. *Philosophy of Right* (translated by T. M. Knox). London: Oxford University Press, 1967.
4 From the chapter on "Ground," in G. W. F. Hegel. *Science of Logic* (translated by A. V. Miller). Atlantic Highlands: Humanities Press International, 1989, 465.
5 Karen Houle. *Responsibility, Complexity, and Abortion: Toward a New Image of Ethical Thought*. Lanham: Lexington Books, 2013: 213.
6 Interview in Jody Duncan. "State of the Art: A Cinefex 25th Anniversary Forum." *Cinefex* 100 (January 2005): 17–107, see 36.
7 Lindsey Buttel blogs indecision, https://prezi.com/qw0akdgwoivz/indecision-hegelian-dialectic-style/ (accessed August 10 2016).
8 Alain Badiou. *Ethics: An Essay on the Understanding of Evil* (translated by Peter Hallward). London: Verso, 2001, 2. In Badiou's *Being and Event*, philosophy begins with a "decision", and there is only one possible decision: "that the one is not" (*BE* 23). The judgment that "the one is" is not a decision at all. The One is imposed, therefore not free, therefore not a decision. In contrast, decision is an excess over the present situation; it cannot be contained in one presentation, it produces a multiplicity and then unfolds into it. Badiou draws his idea of undecidability from Cohen's mathematical proof of the undecidability of the Continuum Hypothesis, "in the absence of any temporality" (*BE* 410). *Being and Event* (translated by Oliver Feltham). London: Continuum, 2005.Under the title "Decision of an Undecidable," (*BE* 405), Badiou describes how a decision, brought about in a subjective act, undoes one situation and affirms a new one. Nothing in a given situation can determine whether or how the decision that undoes it will succeed in creating something new. The means and steps for following a decision through are therefore undecidable. This is first order undecidability of the decision. Secondly, because the decision was undecidable when it was initiated, whatever new situation arises from it will preserve that undecidability (*BE* 406). In our terms, a decision cannot be enacted as a new fact, it can only be enacted as a new decision, or as the same decision to be re-decided. All decisions contain ambiguity, and ambiguity is passed on, so the future of a decision is a series of decisions about how to manage the undecidable terms in it.

9   See Peter Hallward. *Badiou: A Subject to Truth*.
    Minneapolis: University of Minnesota Press, 2003. "The Primacy of
    Decision", 312–315.
10 See Alessandro Carrera. "The Transcendental Limits of Politics: On
    Massimo Cacciari's Political Philosophy," in *Contemporary Italian
    Political Philosophy* (edited by Antonio Calcagno). New York: SUNY
    Series in Contemporary Italian Philosophy, 2015, 119–138. Calcagno
    has played an important role in spotlighting and analyzing these
    philosophers.
11 Charles Petzold. *The Annotated Turing: A Guided Tour through Alan
    Turing's Historic Paper on Computability and the Turing Machine*.
    Indianapolis: Wiley Publishers, 2008.
12 See Paul Livingston's *The Politics of Logic: Badiou, Wittgenstein, and
    the Consequences of Formalism*. New York: Routledge, 2012. On
    Turing, see 151–153.
13 See Leslie Hill. *Radical Indecision: Barthes, Blanchot, Derrida, and
    the History of Criticism*. Notre Dame: University of Notre Dame
    Press, 2010.
14 Jacques Derrida. *On Cosmopolitanism and Forgiveness* (translated
    by Mark Dolley and Michael Hughes). London: Routledge, 2001: 56.
15 Jacques Derrida. *Limited Inc* (translated by Samuel Weber).
    Evanston: Northwestern University Press, 1988.
16 Gary Madison argues that meanings are not so much undecidable
    as inexhaustible. G. B. Madison. *The Hermeneutics of
    Postmodernity: Figures and Themes*. Bloomington: Indiana University
    Press, 1990: 115.
17 Jacques Derrida. *Politics of Friendship* (translated by George Collins).
    London: Verso, [1994] 1997.
18 Jacques Derrida. *The Gift of Death* (translated by David Willis).
    Chicago: University of Chicago Press, [1992] 1995.

9 Deleuze: Decision in the empty future. The virtual decision = X

1   Gilles Deleuze. *Difference and Repetition* (translated by Paul Patton).
    New York: Columbia University Press, 1994. *Différence et repetition*.
    Hereafter "*DR*". Paris: Presses Universitaires de France, 1968.
    I generally refer to page numbers in the translation, but when I give
    my own translations, I refer to page numbers in both texts.
2   Gilles Deleuze and Felix Guattari. *What is Philosophy?* (translated
    by Hugh Tomlinson and Graham Burchell). New York: Columbia
    University Press, 1996: 85–116.

3   Gilles Deleuze. *Cinema 1: The Movement-Image* (translated by Hugh Tomlinson and Barbara Habberjam). Minneapolis: University of Minnesota Press, 1986.
4   For detailed analysis of Deleuze's three syntheses of time, see Chapters 2–4 of my *Deleuze and Guattari's Philosophy of History*. London: Continuum Books, 2006: 12–70. The first portion of this chapter is developed from my "Problems With the Future: Deleuze's *Difference and Repetition*," forthcoming in *Deleuze Studies*, 2018.
5   For example, Henry Somers-Hall. *Deleuze's Difference and Repetition*. Edinburg University Press, 2013: 72–82.
6   Hölderlin (bilingual German/French edition with translation by François Fédier). *Remarques sur Oedipe/Remarques sur Antigone*. France: Union Générale d'Éditions, 1965. Preface by Jean Beaufret, "Hölderlin et Sophocle," 7–42.
7   Daniela Voss connects Hölderlin's caesura with Dedekind's cut. Daniela Voss, "Deleuze's Third Synthesis of Time." *Deleuze Studies* 7, 2 (2013): 194–216.
8   Hölderlin talks about a caesura within a single line of poetry, but also about the caesura in a play when a prophecy announces that things will not continue as they have been. In Sophocles' *Oedipus Rex*, this occurs near the end of the play, changing the storyline to such an extent that the issues at the beginning of the play are in danger of getting lost, and need to be protected. In *Antigone*, the caesura occurs near the beginning, and the end goes so "fast" that nothing much happens later, so the end is in danger of getting lost, and it is what needs to be protected. The cut of time is not in favor of a new and different future, but shows that the situation is hopeless, and the problems that seemed to motivate the action are simply unsolvable. The caesura indicates Antigone's decision to leave this world, and in that way, indicates the end of time.
9   "The libido loses all mnemonic content and time loses its circular shape" (*DR* 113).
10 In *A Thousand Plateaus* (*ATP* 121–4), Deleuze and Guattari envision a straight line of flight that runs away from the city into a hot death in the desert, away from signifiers into a sign-desert.
11 The section on psychoanalysis began with a sharp typographical line break, but it peters out, with the crack about the end of the story of time, a page and a half before the next line break. The line break in the French edition is not marked in the English translation.
12 For analysis of this passage, see my *Simultaneity and Delay*, 166–169.
13 Arkady Plotnitsky. "The Calculable Law of Tragic Representation and the Unthinkable: Rhythm, Caesura and Time, from Hölderlin to Deleuze," in *At the Edges of Thought: Deleuze and Post-Kantian Philosophy* (edited by Craig Lundy and Daniela Voss), Edinburgh: Edinburgh University Press, 2015: 123–145.

14 Émile Benveniste. "*Le langage et l'expérience humaine,*" in *Problèmes de linguistique générale,* 2. Paris: Gallimard, 1974, 76. Deleuze refers to other Benveniste essays from this collection, although not this one.
15 There is a different version of this part of the chapter in my essay "Deleuze's "Power of Decision," Kant's =X, and Husserl's Noema," in *At the Edges of Thought: Deleuze and Post-Kantian Thought* (edited by Craig Lundy and Daniela Voss). Edinburgh: Edinburgh University Press, 2015: 272–292.
16 Gilles Deleuze. "To Have Done with Judgment," in *Essays Critical and Clinical* (translated by Daniel W. Smith and Michael A. Greco). Minneapolis: University of Minnesota Press, 1997: 126–135, especially 134. Deleuze cites D. H. Lawrence.
17 Gilles Deleuze. *Difference and Repetition* (translated by Paul Patton). New York: Columbia University Press, 1994.
18 In *Cinema 2,* Deleuze's use of the term "decision" is again complex and ambivalent. Deleuze complains about the era when "decision-makers" (*décideurs*), bureaucratic meddlers, interfered with artists like Truffaut by controlling film stock. Here he is discussing Serge Daney's idea that cinema might be able to "defeat Hitler cinematographically." Cinema's path to victory would be to create a liberating use of anachronism, to counteract Hitler's fascist use of anachronistic mythology. Whether this is good politics or not, it makes the "decider" look not as bad as the fascist "leader" (*chef*). But there is also a third level of decision-making: in the same passage, Deleuze contrasts the natural organic eye with Vertov's "camera eye," and then with Kubrick's "brain-city." Brain cinema conceives perception as a series of images connected by probabilistic synapse-like intervals. Deleuze goes as far here as he ever does to promote "a new computer and cybernetic race, automata of computation and thought, automata with controls and feedback" (*Cinema 2* 264–5/ 346). Given this context, his reference to "decision-makers" seems to prophesize advances in CGI filmmaking that could only have been dreamed of when *Cinema 2* was published in 1985. For example, Deleuze discusses omni-directional images with no outside images, whose right and left have a "power to turn back on themselves," and other features that sound like *Photoshop* (first released in 1988). In the context of digital cinema, motion graphics, visual effects compositing, AI zombies, and the vast systems of image-control programs, Deleuze's reference to "control and feedback" starts to sound like the power of every image-maker with a laptop to be a "*décideur*" in the new brain city. Gilles Deleuze. *Cinema 2: The Time-Image* (translated by Hugh Tomlinson and Robert Galtea). Minneapolis: University of Minnesota Press, [1985] 1994. *Cinema*

2: *L'Image-temps*. Paris: Les Éditions de Minuit, 1985. See my paper, "Visual Effects and Phenomenology of Perceptual Control." *Cinema: Journal of Philosophy and the Moving Image* 6 (2014): 30–51. http://cjpmi.ifilnova.pt/6-contents

19 Gilles Deleuze. *Logic of Sense* (translated by Constantin V. Boundas). New York: Columbia University Press, [1969] 1990. *Logique du sens*. Paris: Les Éditions de Minuit, 1969. Hereafter "*LS*".

20 Gilles Deleuze. *Foucault* (translated by Sean Hand). Minneapolis: University of Minnesota Press, [1986] 1988.

21 Both Borges stories are in Jorge Luis Borges. *Labyrinths* (translated by Donald A. Yates and John M. Fein). New York: New Directions Publishing, 1962.

22 Immanuel Kant. *Critique of Pure Reason* (translated by Norman Kemp Smith). New York: St. Martin's Press, 1965.

23 Deleuze likewise says that the simple X refers to an unspecified object as an identity. That is different from the = X, "the unformed paradoxical element which ... misses its own identity" (*LS* 119 and 145). Deleuze's vocabulary is not always identical. In one passage, Deleuze distinguishes the object = X from the thing = X (*LS* 145). The former he defines as a commonsense identification of an object with many properties (*LS* 228). This is more like Kant's "object X." In this passage, the "thing = X" is defined as the paradoxical constellation without identity, which is more like the object = X in other passages.

24 Gilles Deleuze. *Kant's Critical Philosophy* (translated by Hugh Tomlinson and Barbara Habberjam). Minneapolis: University of Minnesota Press, 1996.

25 Béatrice Longuenesse. *Kant and the Capacity to Judge* (translated by Charles T. Wolfe). Princeton: Princeton University Press, 1998: 48–49.

26 Alain Badiou. "L'ontologie soustractive de Kant," in *Court Traité d'ontologie transitoire*. Paris: Éditions de Seuil, 1998: 153–164.

27 Philip McPherson Rudisill. "Circles in the Air: Pantomimics and the Transcendental Object=X." *Kant-Studien* 87, 2 (1996): 132–148.

28 Martin Heidegger. *Kant and the Problem of Metaphysics* (translated by Richard Taft). Bloomington: Indiana University Press, 1997.

29 Juliette Simont. *Essai sur la quantité, la qualité, la relation chez Kant, Hegel, Deleuze: Les "fleurs noires" de la logique philosophique.* Paris: Éditions L'Harmattan, 1997.

30 Alain Beaulieu. *Gilles Deleuze et la phénoménologie*. Mons, Belgique: Les Editions Sols Maria, 2004.

31 Gilles Deleuze. Cours Vincennes – 28/03/1978, at webdeleuze.com [1978] (accessed January 2013).

32 See Elhanan Yakira. "Salomon Maimon and the Question of Predication," in *Salomon Maimon: Rational Dogmatist, Empirical Skeptic: Critical Assessments* (edited by G. Freudenthal). New York: Springer, 2003: 54–79.
33 Daniel Smith applies Deleuze's lecture to Bacon's painting. "Deleuze on Bacon: Three Conceptual Trajectories in 'The Logic of Sensation,'" in *Essays on Deleuze* (edited by Daniel W. Smith). Edinburgh: Edinburgh University Press, 2012: 222–234.
34 Johann Gottlieb Fichte, *Science of Knowledge* (translated by Peter Heath and John Lachs). New York: Appleton-Century-Crofts, 1970.
35 Johann Gottlieb Fichte, *Science of Right* (translated by Adolph Ernst Kroeger). Philadelphia: J. B. Lippincott, 1869.
36 In the same passage, Deleuze associates the "object = X" with the phallus (Deleuze *LS* 233). An action = X names a place of action where it will never be found present, just as a phallus points, or as a mother hides (*DR* 103 and 105). To tell the story straight up: An ego desires an ideal, distributes its action in practical images over time, but never completely fills time. To tell the story in Freudian terms (*DR* 110): The narcissist desires to find its ego ideal, but fails when confronted by the whole of time that is too big for it. The whole of time is predicted, but also forbidden, by the superego. The ego finds time again only if it de-sexualizes its energies, sublimating its grand ideals in practical effects, unravelling the endless cycles of time, and satisfying itself on linear time. In "*À quoi reconnaît-on le structuralisme?*" (in *L'Île déserte et autres textes*. Paris: Les Éditions de Minuit, 2002: 238–269), Deleuze brings together many of the definitions of = X, starting with psychoanalytic structuralism and ending with a loose connection between = X and decision:

1. The = X is the "empty square" on all series at once, given different meanings by whoever encounters it (like the handkerchief in *Othello*, Deleuze *À Quoi* 258–9).
2. = X is like the dummy hand in bridge (Deleuze cites Lacan, *À Quoi* 259).
3. = X is like the phallus, not an object, but an exchange (Deleuze *À Quoi* 263).
4. = X is a *perpetuum mobile* (Deleuze *À Quoi* 262).
5. = X is nevertheless not unintelligible or indeterminate (Deleuze *À Quoi* 265).

Edward Willatt connects the structuralist empty square = X with Deleuze's "strange object." "The Genesis of Cognition: Deleuze as a Reader of Kant," in *Thinking Between Deleuze and Kant* (edited

by Edward Willatt and Matt Lee). London: Continuum Books, 2009: 67–85.
37 Gilles Deleuze. *Nietzsche and Philosophy* (translated by Hugh Tomlinson). New York: Columbia University Press, [1962] 1983. Chapter 1, section 11, "The Dicethrow" and surrounding sections, 24–31.
38 Edmund Husserl. *Ideas Pertaining to a Pure Phenomenology and to a Phenomenological Philosophy, First Book* (translated by Fred Kersten). Dordrecht: Kluwer Academic Publishers, 1982.
39 Edmund Husserl. *Analyses Concerning Passive and Active Synthesis* (translated by Anthony J. Steinbock). Dordrecht: Kluwer Academic Publishers, 2001.
40 One might imagine decision-making games where decisions of other sorts are found. To take a surrealist example, admittedly with not many useful applications, Raul Ruiz imagines a game where herds of cunning and fearful dice make decisions for themselves. Raul Ruiz. *Poetics of Cinema 1* (translated by Brian Holmes). Paris: Éditions Dis Voir, 2005: 19–20.
41 There may exist entirely chance forces, but problems, decisions, and games are not among them. My view is that chance effects constitute limited parts of machinic assemblages.
42 How do many thoughts "communicate in one long thought, causing all the forms or figures of the nomadic distribution to correspond to its own displacement, everywhere insinuating chance and ramifying each thought, linking the 'once and for all' to 'each time' for the sake of 'all time' " (Deleuze *Nietzsche* 60)?
43 Gilles Deleuze. *Logique du sens*. Paris: Les Éditions de Minuit, 1969. My translation: "It behooves the singular points to be distributed according to mobile and communicating figures which make of every dice throw (*coups*) one and the same cast (*lancer*) and of this cast a multiplicity of throws (*coups*)."
44 François Truffaut. *The Films in my Life* (translated by Leonard Mayhew). New York: Touchstone Books, 1985: 152–154. Jean-Luc Godard. *Godard on Godard* (translated by Tom Milne). New York: Da Capo Press, 1972, 57–59.
45 Gilles Deleuze. *The Fold* (translated by Tom Conley). Minneapolis: University of Minnesota Press, 1993.
46 David Woodruff Smith and Ronald McIntyre. *Husserl and Intentionality*. Dordrecht: Reidel Publishing Company, 1982: 292–295.
47 David Lewis. *On the Plurality of Worlds*. Malden: Blackwell Publishing, 1986. When confronted with problems like those I raise, Lewis tends to respond: "Who cares?"

48 There are many elements of Leibniz's possible worlds that Deleuze does not side with. For one thing, Leibniz does not allow incompossible worlds to "exist at once" (*Fold* 62). For another, Leibniz is unclear whether every small discernable difference designates a discrete substance, or whether there is a continuity of infinitely small differences within any event (*Fold* 65). This in turn makes it unclear what exactly an individual includes, and what ethical responsibilities an individual has to vary her properties. On the other hand, there are points where Deleuze may follow Leibniz too closely: It is partly to conform to Leibniz's theory that monads are wholes, and do not evolve, that Deleuze's descriptions of divergence do not sound much like projections into the future.
49 Gottfried Wilhelm Leibniz. *Annotatio de quibusdam Ludis; inprimis de Ludo quodam Sinico, differentiaque Scachici et Latrunculorum, et novo genere Ludi Navalis*. In *Miscellanea Berolinesia* 1 (1710): 22–26.
50 For definitions, examples, and controversies, see "Tewari", http://senseis.xmp.net/?Tewari (accessed April 27 2017).
51 *The New English Bible*. New York: Oxford University Press, 1972.

Afterword

1 William Gibson. *Distrust that Particular Flavor*. New York: Penguin, 2012.
2 Sun Ra. *Space is the Place*. New York: ABC Records, 1973.
3 Quoted in Miyamoto Naoki. "Annals of Handicap Go: Sekiyama v. Miki (2)." *Go World* Magazine, Tokyo: Ishi Press, No. 12, 1979: 51.

BIBLIOGRAPHY

Adorno, Theodor W. *Negative Dialectics* (translated by E. B. Ashton). New York: Continuum, 1973.
Agamben, Giorgio. *Homo Sacer: Sovereign Power and Bare Life* (translated by Daniel Heller-Roazen.). Palo Alto: Stanford University Press, 1998.
Anscombe, G. E. M. *Intentions*. Ithaca: Cornell University Press, 1957.
Aristotle, *De Interpretatione* (translated by E. M. Edghill), in *The Basic Works of Aristotle* (edited by Richard McKeon). New York: Random House, 1941.
Aristotle, *Ethics* (translated by W. D. Ross), in *The Basic Works of Aristotle* (edited by Richard McKeon). New York: Random House, 1941.
Austin, J. L. *How to do Things with Words*. Cambridge: Harvard University Press, [1962] 1975.
Badiou, Alain. *Being and Event* (translated by Oliver Feltham). London: Continuum, [1988] 2005.
Badiou, Alain. *Ethics: An Essay on the Understanding of Evil* (translated by Peter Hallward). London: Verso, 2001.
Badiou, Alain. "L'ontologie soustractive de Kant," in *Court Traité d'ontologie transitoire*. Paris: Éditions de Seuil, 1998: 153–64.
Badiou, Alain. *Number and Numbers* (translated by Robin Mackey). Cambridge: Polity Press, 2008.
Baho, James Scott. *Heidegger and Deleuze: The Groundwork of Evental Ontology*. PhD Dissertation, Duquesne University, 2016.
Balibar, Étienne. "Sur les concepts fondamentaux du matérialisme historique," in *Lire le Capital* (edited by Louis Althusser et al.). Paris: Presses Universitaires de France, 1965: 419–568.
Bartol, Jordan and Stefan Linquist. "How do Somatic Markers Feature in Decision Making?" *Emotion Review* 7,1 (2015): 81–89.
Bates, Jennifer Ann. "Hamlet and Kierkegaard on Outwitting Recollection," in *Shakespeare and Continental Philosophy* (edited by Jennifer Ann Bates and Richard Wilson). Edinburgh: Edinburgh University Press, 2014: 40–55.

Baugh, Bruce. "Freedom, Fatalism and the Other in *Being and Nothingness* and *The Imaginary*," in *Phenomenology 2010*, Volume 5 (edited by Lester Embree, Michael Barber, and Thomas Nenon). Bucharest: Zenon Books, 2010: 199–218.

Beaulieu, Alain. *Gilles Deleuze et la phénoménologie*. Mons, Belgique: Les Éditions Sols Maria, 2004.

Belknap, Nuel and Michael Perloff. "Seeing to it that: A Canonical Form for Agentives." *Theoria* 54 (1988): 175–199.

Benjamin, Walter. "Critique of Violence," in *Reflections* (edited by Peter Demetz, translated by Edmund Jephcott). New York: Schocken, 1986: 277–300.

Benveniste, Émile. "Le langage et l'expérience humaine," in *Problèmes de linguistique générale*, 2. Paris: Gallimard, 1974.

Bergson, Henri. *Duration and Simultaneity* (translated by Mark Lewis and Robin Durie). Manchester: Clinamen Press, [1922] 1999.

Bergson, Henri. *Matter and Memory* (translated by N. M. Paul and W. S. Palmer). New York: Zone Books, [1896] 1991.

Berliner, Paul F. *Thinking in Jazz: The Infinite Art of Improvisation*. Chicago: University of Chicago Press, 1994.

Berthoz, Alain. *Reason and Emotion: The Cognitive Neuroscience of Decision Making* (translated by Giselle Weisse). Oxford: Oxford University Press, 2006.

Borges, Jorge Luis Borges. *Labyrinths* (translated by Donald A. Yates and John M. Fein). New York: New Directions Publishing, 1962.

Boulé, Jean-Pierre. *Sartre, Self-Formation and Masculinities*. New York: Berghahn Books, 2005.

Braddon-Mitchell, David and Caroline West. "Temporal Phase Pluralism." *Philosophy and Phenomenological Research* 62,1 (January 2001): 59–83.

Briggs, Rachael and Graeme Forbes. "The Real Truth about the Unreal Future," in *Oxford Studies in Metaphysics, V 7* (edited by Karen Bennett and Dean W. Zimmerman). Oxford: Oxford University Press, 2012: 257–304.

Broad, C. D. *Scientific Thought*. London: Kegan Paul, Trench, Trubner, 1923.

Bruin, John. *Homo Interrogans: Questioning and the Intentional Structure of Cognition*. Ottawa: University of Ottawa Press, 2001.

Brunner, John. *Timescoop*. New York: Dell, 1969.

Bunzel, Dirk. "The Rhythm of the Organization: Simultaneity, Identity, and Discipline in an Australian Coastal Hotel," in *Making Time: Time and Management in Modern Organizations* (edited by Richard Whipp, Barbara Adam, and Ida Sabeus). Oxford: Oxford University Press, 2002: 168–181.

Burch, Matthew. "Death and Deliberation: Overcoming the Decisionism Critique of Heidegger's Practical Philosophy." *Inquiry*, 53, 3 (June 2010): 211–234.

Butler, Judith. *Precarious Life: The Powers of Mourning and Violence.* London: Verso, 2006.

Buttel, Lindsey. [No title], https://prezi.com/qw0akdgwoivz/indecision-hegelian-dialectic-style/ (accessed August 10 2016).

Cardozo, Benjamin N. *The Nature of the Judicial Process.* Lexington: CreateSpace Independent Publishing Platform, [1921] 2012.

Carrera, Alessandro. "The Transcendental Limits of Politics: On Massimo Cacciari's Political Philosophy," in *Contemporary Italian Political Philosophy* (edited by Antonio Calcagno). New York: SUNY Series in Contemporary Italian Philosophy, 2015: 119–138.

Center for Technology and Behavioral Health, Dartmouth College. http://www.c4tbh.org/the-center/what-we-re-up-to/active-projects/flexible-decision-making-in-response-to-unplanned-events (accessed September 20 2014).

Chrétien, Jean-Louis. *The Unforgettable and the Unhoped For* (translated by Jeffrey Bloechl). New York: Fordham University Press, 2002.

Chu, John. "Thirty Seconds from Now." in *The Time Traveler's Almanac* (edited by Ann and Jeff VanderMeer). New York: Tor, 2014: 215–221. Originally published in *Boston Review*, 2011.

Cunningham, Arthur J. "Branches in the Everett Interpretation." *Studies in the History and Philosophy of Science Part B: Studies in the History and Philosophy of Modern Physics.* 46, Part B, (May 2014): 247–262.

Currie, Gregory. *Image and Mind: Film, Philosophy, and Cognitive Science.* Cambridge: Cambridge University Press, 1995.

Curtis, Adam. *All Watched Over by Machines of Loving Grace* [Film]. BBC Productions, 2011.

De Beauvoir, Simone. *The Ethics of Ambiguity* (translated by Bernard Frechtman). New York: Citadel Press, [1947] 1991.

Deleuze, Gilles. "À quoi reconnaît-on le structuralisme?" in *L'Île déserte et autres textes.* Paris: Les Éditions de Minuit, [1972] 2002: 238–269.

Deleuze, Gilles. *Cinema 1: The Movement-Image* (translated by Hugh Tomlinson and Barbara Habberjam). 1980. Minneapolis: University of Minnesota Press, [1983] 1986.

Deleuze, Gilles. *Cinema 2: The Time-Image* (translated by Hugh Tomlinson and Robert Galtea). 1985. Minneapolis: University of Minnesota Press, [1985] 1994. *Cinema 2: L'Image-temps.* Paris: Les Éditions de Minuit.

Deleuze, Gilles. *Difference and Repetition* (translated by Paul Patton). New York: Columbia University Press, [1968] 1994. *Différence et repetition.* Paris: Presses Universitaires de France.

Deleuze, Gilles. Cours Vincennes – 28/03/1978, at webdeleuze.com [1978] (accessed January 2013).
Deleuze, Gilles. *Foucault* (translated by Sean Hand). Minneapolis: University of Minnesota Press, [1986] 1988.
Deleuze, Gilles. *Nietzsche and Philosophy* (translated by Hugh Tomlinson). New York: Columbia University Press, [1962] 1983.
Deleuze, Gilles. *Kant's Critical Philosophy* (translated by Hugh Tomlinson and Barbara Habberjam). Minneapolis: University of Minnesota Press, [1963] 1996.
Deleuze, Gilles. *Logic of Sense* (translated by Constantin V. Boundas). New York: Columbia University Press, [1969] 1990. *Logique du sens*. Paris: *Les Éditions de Minuit*.
Deleuze, Gilles. *The Fold* (translated by Tom Conley). Minneapolis: University of Minnesota Press, [1988] 1993.
Deleuze, Gilles. "To Have Done with Judgment," in *Essays Critical and Clinical* (translated by Daniel W. Smith and Michael A. Greco). Minneapolis: University of Minnesota Press, 1997: 126–135.
Deleuze, Gilles and Guattari, Felix. *Anti-Oedipus* (translated by Robert Hurley, Mark Seem, and Helen R. Lane). Minneapolis: University of Minnesota Press, [1972] 1983.
Deleuze, Gilles. *A Thousand Plateaus* (translated by Brian Massumi). Minneapolis: University of Minnesota Press, [1980] 1987.
Deleuze, Gilles. *What is Philosophy?* (translated by Hugh Tomlinson and Graham Burchell). New York: Columbia University Press, [1990] 1996.
Denis, Benoît. "*Retards de Sartre*", *Études Sartriennes*, 2005, no. 10: 189–209.
Derrida, Jacques. *Limited Inc* (translated by Samuel Weber). Evanston: Northwestern University Press, 1988.
Derrida, Jacques. *Politics of Friendship* (translated by George Collins). London: Verso, 1997.
Derrida, Jacques. *On Cosmopolitanism and Forgiveness* (translated by Mark Dooley and Michael Hughes). London: Routledge, 2001.
Derrida, Jacques. *The Gift of Death* (translated by David Willis). Chicago: University of Chicago Press, 1995.
Des Jardins, Joseph R. *Environmental Ethics*. Belmont: Wadsworth, 2012.
Deutsch, David. "Interview," in *The Ghost in the Atom: A Discussion of the Mysteries of Quantum Physics* (edited by P. C. W. Davies and J. R. Brown). Cambridge: Cambridge University Press, 1986: 83–105.
Deutsch, David. *The Fabric of Reality: The Science of Parallel Universes—and its Implications*. New York: Penguin Press, 1997.
Drummond, John J. *Husserlian Intentionality and Non-Foundational Realism: Noema and Object*. Dordrecht: Kluwer Academic Publishers, 1990.

Dummett, Michael. "The Reality of the Past," in *Truth and Other Enigmas*. Cambridge: Harvard University Press, 1978: 358–374.

Duncan, Jody. "State of the Art: A *Cinefex* 25th Anniversary Forum." *Cinefex* 100, (January 2005): 17–107.

Esposito, Elena. "The Hypertrophy of Simultaneity in Telematic Communication." *Thesis Eleven* 51 (1998): 17–36.

Eye in the Sky, DVD. Directed by Gavin Hood. Hollywood: Raindog Films, 2015.

Fichte, Johann Gottlieb. *Science of Knowledge* (translated by Peter Heath and John Lachs). New York: Appleton-Century-Crofts, [1794] 1970.

Fichte, Johann Gottlieb. *Science of Right* (translated by Adolph Ernst Kroeger). Philadelphia: J. B. Lippincott, [1797] 1869.

Forrester, Jay W. "Counterintuitive Behavior of Social Systems." *Technology Review* 73, 3 (1971): 52–68.

Forrester, Jay W. *Urban Dynamics*. Cambridge: MIT. Press, 1969.

Foton, Nick. "Introduction," in *Contingent Future Persons: On the Ethics of Deciding Who Will Live, or Not, in the Future* (edited by Nick Foton and Jan C. Heller). Dordrecht: Kluwer Academic Publishers, 1997: 1–8.

Fox, Justin. *The Myth of the Rational Market: A History of Risk, Reward, and Delusion on Wall Street*. New York: Harper Collins, 2009.

Gell-Mann, Murray. *The Quark and the Jaguar: Adventures in the Simple and the Complex*. New York: W. H. Freeman, 1994.

Gerrans, Philip. "Mental time travel, somatic markers and 'myopia for the future.'" *Synthese* 159 (2007): 459–474.

Gibson, William. *Distrust that Particular Flavor*. New York: Penguin, 2012.

Godard, Jean-Luc. *Godard on Godard* (translated by Tom Milne). New York: Da Capo Press, 1972.

Habermas, Jürgen. *Between Facts and Norms* (translated by William Rehg). Cambridge: MIT Press, [1992] 1998.

Habermas, Jürgen. *Legitimation Crisis* (translated by Thomas McCarthy). Boston: Beacon Press, [1973] 1975.

Haldeman, Joe. *The Forever War*. New York: Ballantine Books, 1974.

Hallward, Peter. *Badiou: A Subject to Truth*. Minneapolis: University of Minnesota Press, 2003.

Hamilton, Andy. "Intention and the Authority of Avowals." *Philosophical Explorations* 11, 1 (March 2008): 23–37.

Hansel and Gretel: Witch Hunters, DVD. Directed by Tommy Wirkola. Hollywood: Paramount Studios, 2013.

Heap, Shaun P. Hargreaves and Yanis Varoufakis, *Game Theory: A Critical Introduction*. London: Routledge, 1995.

Hegel, G. W. F. *Phenomenology of Spirit* (translated by A. V. Miller). Oxford: Oxford University Press, [1807] 1977. *Phänomenologie des Geistes*. Frankfurt: Suhrkamp, 1973.

Hegel, G. W. F. *Philosophy of Right* (translated by T. M. Knox). London: Oxford University Press, [1820] 1967.

Hegel, G. W. F. *Science of Logic* (translated by A. V. Miller). Atlantic Highlands: Humanities Press International, [1816] 1989.

Hegel, G. W. F. *The Difference Between Fichte's and Schelling's Systems of Philosophy* (translated by H. S. Harris and Walter Cerf). New York: SUNY Press, [1801] 1998.

Heidegger, Martin. *Being and Time* (translated by John Macquarrie and Edward Robinson). NY: Harper and Row, Publishers, [1927] 1962.

Heidegger, Martin. *Contributions to Philosophy (From Enowning)* (translated by Parvis Emad and Kenneth Maly). Bloomington: Indiana University Press, [1937] 1999. *Beiträge zur Philosophie (vom Ereignis). Gesamtausgabe, Band 65*. Frankfurt am Main: Vittorio Klostermann, 1989.

Heidegger, Martin. *Introduction to Metaphysics* (translated by Gregory Fried and Richard Polt). New Haven: Yale University Press, [1935] 2000.

Heidegger, Martin. *Kant and the Problem of Metaphysics* (translated by Richard Taft). Bloomington: Indiana University Press, [1929] 1997.

Henry, Claude. "Investment Decisions Under Uncertainty: The 'Irreversibility Effect.'" *The American Economic Review* 64, 6 (1974): 1006–1012.

Henry, Michel. *Phénoménologie matérielle*. Paris: Presses Universitaires de France, 1990.

Hill, Leslie. *Radical Indecision: Barthes, Blanchot, Derrida, and the History of Criticism*. Notre Dame: University of Notre Dame Press, 2010.

Hölderlin. *Remarques sur Oedipe/Remarques sur Antigone* (bilingual German/French edition with translation by François Fédier). France: Union Générale d'Éditions, [1804] 1965. Preface by Jean Beaufret, "Hölderlin et Sophocle", 7–42.

Horwich, Paul. *Asymmetries in Time: Problems in the Philosophy of Science*. Cambridge: MIT Press, 1987.

Houle, Karen. *Responsibility, Complexity, and Abortion: Toward a New Image of Ethical Thought*. Lanham: Lexington Books, 2013.

Hume, David. *David Hume's Political Essays* (edited by Charles W. Hendel). Indianapolis: Bobbs-Merrill, 1953.

Husserl, Edmund. *Analyses Concerning Passive and Active Synthesis* (translated by Anthony J. Steinbock). Dordrecht: Kluwer Academic Publishers, [1926] 2001.

Husserl, Edmund. *Experience and Judgment* (edited by Ludwig Landgrebe, translated by James S. Churchill and Karl Ameriks).

Evanston: Northwestern University Press, [1939] 1973. *Erfahrung und Urteil.* Hamburg: Felix Meiner Verlag, 2014.

Husserl, Edmund. *Ideas Pertaining to a Pure Phenomenology and to a Phenomenological Philosophy, First Book* (translated by Fred Kersten). Dordrecht: Kluwer Academic Publishers, [1913] 1982. *Ideen zu einer reinen Phänomenologie und phänomenologischen Philosophie.* Tübingen: Max Niemeyer Verlag, 1993.

James, William. "Draft on Brain Procedures and Feelings," in *Manuscript Essays and Notes* (General Editor: Frederick H. Burkhardt). Cambridge: Harvard University Press, [1872] 1988, 247–255.

James, William. *The Principles of Psychology*, V. 2. Mineola: Dover Publications, [1890] 1950.

James, William. "The Will to Believe," in *The Will to Believe and Other Essays in Popular Philosophy.* Mineola: Dover, [1896] 1956: 1–31.

Jankélévitch, Vladimir. *Henri Bergson.* Paris: Presses Universitaires de France, 1989.

Jarry, Alfred. *Exploits and Opinions of Dr. Faustroll, Pataphysician* (translated by Simon Watson Taylor.) Boston: Exact Change, [1898] 1996.

Jaspers, Karl. *Philosophy* V. 2 (translated by E. B. Ashton). Chicago: University of Chicago Press, [1932] 1970.

Johannesen, Jon-Arild, Johan Olaisen, and Bjørn Olsen. "The Philosophy of Science, Planning, and Decision Theories." *Built Environment* 24, 2/3 (1998): 155–168.

Kahneman, Daniel. *Thinking, Fast and Slow.* New York: Farrar, Straus and Giroux, 2011.

Kalyvas, Andreas. "From the Act to the Decision: Hannah Arendt and the Question of Decisionism." *Political Theory* 32, 3 (June 2004): 320–346.

Kant, Immanuel. *Critique of Pure Reason* (translated by Norman Kemp Smith). New York: St. Martin's Press, [1781] 1965.

Kant, Immanuel. *Critique of Practical Reason* (translated by Lewis White Beck). Indianapolis: Bobbs-Merrill, [1788] 1956.

Keynes, John Maynard. 1936. *The General Theory of Employment, Interest and Money.* Cambridge: Cambridge University Press. Reprinted by BN Publishing, 2008.

Kierkegaard, Søren. *Concluding Unscientific Postscript*, V. 1 (translated by Howard V. Hong and Edna H. Hong). Princeton: Princeton University Press, [1846] 1992.

Kierkegaard, Søren. *Philosophical Fragments* (translated by Howard V. Hong and Edna H. Hong). Princeton: Princeton University Press, [1844] 1985.

Lampert, Jay. *Deleuze and Guattari's Philosophy of History*. London: Continuum Books, 2006.
Lampert, Jay. "Deleuze's "Power of Decision", Kant's =X, and Husserl's Noema," in *At the Edges of Thought: Deleuze and Post-Kantian Thought* (edited by Craig Lundy and Daniela Voss). Edinburgh: Edinburgh University Press, 2015: 272–292.
Lampert, Jay. "Problems With the Future: Deleuze's *Difference and Repetition*." Forthcoming in *Deleuze Studies*, 2018.
Lampert, Jay. *Simultaneity and Delay*. London: Continuum Books, 2012.
Lampert, Jay. "Visual Effects and Phenomenology of Perceptual Control." *Cinema: Journal of Philosophy and the Moving Image* 6 (2014): 30–51. http://cjpmi.ifilnova.pt/6-contents.
Leibniz, Gottfried Wilhelm. *Annotatio de quibusdam Ludis; inprimis de Ludo quodam Sinico, differentiaque Scachici et Latrunculorum, et novo genere Ludi Navalis*. *Miscellanea Berolinesia* 1 (1710): 22–26.
Leibniz, Gottfried Wilhelm. "On the Ultimate Origination of Things" (edited and translated by Daniel Garber and Roger Ariew), in *Discourse on Metaphysics and Other Essays*. Indianapolis: Hackett Publishing, [1697] 1991: 41–48.
Lewis, David. *On the Plurality of Worlds*. Malden: Blackwell Publishing, 1986.
Lichtheim, Miriam (editor and translator). *Ancient Egyptian Literature: A Book of Readings. Volume 1: The Old and Middle Kingdoms*. Berkeley: University of California Press, 1975.
Linz, Juan J. "Democracy's Time Constraints." *International Political Science Review* 19, 1 (1998): 19–37.
Livingston, Paul. *The Politics of Logic: Badiou, Wittgenstein, and the Consequences of Formalism*. New York: Routledge, 2012.
Lodzinski, Don. "The Eternal Act." *Religious Studies* 34, 3 (Spring 1998): 325–352.
Looper, DVD. Directed by Rian Johnson. Hollywood: Endgame Entertainment, 2012.
Longuenesse, Béatrice. *Kant and the Capacity to Judge* (translated by Charles T. Wolfe). Princeton: Princeton University Press, 1998.
Luhmann, Niklas. *Risk: A Sociological Theory* (translated by Rhodes Barrett). New York: Aldine de Gruyter, 1993.
Lyotard, Jean-François and Jean-Loup Thébaud. *Just Gaming* (translated by Wlad Godzich). Minneapolis: University of Minnesota Press, 1985.
Machiavelli, Niccolò. *The Discourses* (translated by Leslie J. Walker). Middlesex: Penguin Books, 1970.
Madison, Gary Bent. *The Hermeneutics of Postmodernity: Figures and Themes*. Bloomington: Indiana University Press, 1990.

Marion, Jean-Luc. *Being Given: Toward a Phenomenology of Givenness* (translated by Jeffrey L. Kosky). Stanford: Stanford University Press, 2002.

Mbiti, John S. *African Religions and Philosophy*. New York: Anchor Books, 1970.

McCall, Storrs. *A Model of the Universe: Space-Time, Probability, and Decision*. Oxford: Oxford University Press, 1994.

McDermott, Michael. "Are Plans Necessary?" *Philosophical Studies* 138, 2 (March 2008): 225–232.

McTaggart, John McTaggart Ellis. *The Nature of Existence*, V. 2. Cambridge: Cambridge University Press, [1927] 1968.

Meinong, Alexius. *On Assumptions* (translated by James Heanue). Berkeley: University of California Press, 1983.

Meyer, Lukas H. "More Than They Have a Right to: Future People and Our Future Oriented Projects," in *Contingent Future Persons: On the Ethics of Deciding Who Will Live, or Not, in the Future* (edited by Nick Foton and Jan C. Heller). Dordrecht: Kluwer Academic Publishers, 1997: 137–156.

Molina, Luis. *On Divine Foreknowledge: Part IV of the Concordia* (translated by Alfred Freddosa). Ithaca: Cornell University Press, [1588] 1988.

Murchadha, Felix Ó. "Love's Conditions: Passion and the Practice of Philosophy," in *Thinking About Love: Essays in Contemporary Continental Philosophy* (edited by Diane Enns and Antonio Calcagno). University Park: Penn State University Press, 2017: 81–97.

Naoki, Miyamoto. "Annals of Handicap Go: Sekiyama v. Miki (2)." *Go World* Magazine, Tokyo: Ishi Press, No. 12, 1979: 50–51.

Nahin, Paul J. *Time Machines: Time Travel in Physics, Metaphysics, and Science Fiction* (second edition). New York: Springer-Verlag, 1999.

The New English Bible. New York: Oxford University Press, 1972.

Nietzsche, Friedrich. *On the Genealogy of Morals* (translated by Walter Kaufman and R. J. Hollingdale). New York: Vintage Books, [1887] 1969.

Ockham, William. *Predestination, God's Foreknowledge, and Future Contingents* (translated by Marilyn McCord Adams and Norman Kretzmann). New York: Appleton-Century-Crofts, [1324] 1969.

Paikin, Steve. *Public Triumph Private Tragedy: The Double Life of John P. Robarts*. Toronto: Viking Canada, 2005.

Panse, Silke. "The Judging Spectator in the Image," in *A Critique of Judgment in Film and Television* (edited by Silke Panse and Dennis Rothermel). New York: Palgrave Macmillan, 2014: 33–70.

Pascal, Blaise. *Pensées* (translated by H. F. Stewart). New York: Modern Library, [1669] 1967.
Pasolini, Pier Paolo. *Heretical Empiricism* (translated by Ben Lawton and Louise K. Barnett). Bloomington: Indiana University Press, 1988.
Paterson, Katie. *Future Library*. https://www.futurelibrary.no (accessed April 12 2017).
Peris, Daniel. *The Strategic Dividend Investor*. New York: McGraw Hill, 2011.
Peterson, Martin. *An Introduction to Decision Theory*. Cambridge: Cambridge University Press, 2009.
Petzold, Charles. *The Annotated Turing: A Guided Tour through Alan Turing's Historic Paper on Computability and the Turing Machine*. Indianapolis: Wiley, 2008.
Plotnitsky, Arkady. "The Calculable Law of Tragic Representation and the Unthinkable: Rhythm, Caesura and Time, from Hölderlin to Deleuze," in *At the Edges of Thought: Deleuze and Post-Kantian Philosophy* (edited by Craig Lundy and Daniela Voss). Edinburgh: Edinburgh University Press, 2015: 123–145.
Price, Huw. *Time's Arrow and Archimedes' Point: New Directions for the Physics of Time*. Oxford: Oxford University Press, 1997.
Primer, DVD. Directed by Shane Carruth. Hollywood: ERBP Films, 2004.
Randall, Lisa. *Unraveling the Mysteries of the Universe's Hidden Dimensions*. New York: Harper Perennial, 2006.
Rawls, John. *A Theory of Justice*. Cambridge: Harvard University Press, 1971.
Redmond, Michael. Live commentaries of the five games between the Google computer program AlphaGo and the human Lee Sedol 9D, March 9–15, 2016 (Youtube). https://www.youtube.com/watch?v=vFr3K2DORc8 (accessed March 15 2016).
Rommelfanger, H. "Fuzzy Decision Theory" (no year listed). www.wiwi.uni-frankfurt.de/profs/rommelfanger/index/dokumente/Aufsatz%20EOLSS%206.5%202002.pdf (accessed September 2016).
Rousseau, Jean-Jacques. *On the Social Contract* (translated by Donald A. Cress). Indianapolis: Hackett Classics, [1762] 1988.
Rudisill, Philip McPherson. "Circles in the Air: Pantonimics and the Transcendental Object=X." *Kant-Studien* 87, 2 (1996): 132–148.
Ruiz, Raul. *Poetics of Cinema 1* (translated by Brian Holmes). Paris: Éditions Dis Voir, 2005.
Sartre, Jean-Paul. *Being and Nothingness* (translated by Hazel E. Barnes). New York: Philosophical Library, [1945] 1956.
Sartre, Jean-Paul. *Notebooks from a Phony War 1939–40* (translated by Quintin Hoare). London: Verso, [1940] 1999. *Les carnets de la drôle de guerre*. Paris: Gallimard, 1983.

Schelling, Thomas C. *The Strategy of Conflict*. New York: Oxford University Press, 1963.

Schmitt, Carl. *Political Theology: Four Chapters on the Concept of Sovereignty* (translated by George Schwab). Chicago: University of Chicago Press, [1934] 2005.

Schmitter, Philippe C. and Javier Santiso. "Three Temporal Dimensions to the Consolidation of Democracy." *International Political Science Review* 19, 1 (1998): 69–92.

Schrag, Calvin O. *Existence and Freedom: Towards an Ontology of Human Finitude*. Evanston: Northwestern University Press, 1961.

Sicard, Michel. "*La nourriture du philosophe.*" *Études Sartriennes* no. 15 (2011): 61–90.

Silver, David et al. (Google DeepMind). "Mastering the Game of Go with Deep Neural Networks and Tree Search." *Nature* 529 (28 January 2016): 484–489.

Simak, Charles. *Time is the Simplest Thing*. New York: Crest Books, 1961.

Simont, Juliette. *Essai sur la quantité, la qualité, la relation chez Kant, Hegel, Deleuze: Les "fleurs noires" de la logique philosophique*. Paris: Éditions L'Harmattan, 1997.

Smart, J. J. C. "Is Time Travel Possible?." *The Journal of Philosophy* 60, 9 (1963): 237–241.

Smith, Daniel. "Deleuze on Bacon: Three Conceptual Trajectories in 'The Logic of Sensation,' " in *Essays on Deleuze* (edited by Daniel W. Smith). Edinburgh: Edinburgh University Press, 2012: 222–234.

Smith, David Woodruff and Ronald McIntyre. *Husserl and Intentionality*. Dordrecht: Reidel Publishing, 1982.

Sokolowski, Robert. *Eucharistic Presence: A Study in the Theology of Disclosure*. Washington DC: Catholic University of America Press, 1994.

Sokolowski, Robert. *Moral Action: A Phenomenological Study*. Bloomington: Indiana University Press, 1985.

Somers-Hall, Henry. *Deleuze's Difference and Repetition*. Edinburgh University Press, 2013.

Stein, Eric W. "Improvisation as Model for Real-Time Decision Making," in *Supporting Real Time Decision-Making: The Role of Context in Decision Support on the Move* (edited by Frada Burstein, Patrick Brézillon, and Arkady Zaslavsky). New York: Springer, 2010: 13–33.

Sun Ra. *Space is the Place*. New York: ABC Records, 1973.

Szendy, Peter. *Kant in the Land of Extraterrestrials: Cosmopolitical Philosofictions* (translated by Will Bishop). New York: Fordham University Press, 2013.

Taleb, Nassim Nicholas. *The Black Swan: The Impact of the Highly Improbable*. New York: Random House, 2007.

Tetlock, Philip E. and Dan Gardner. *Superforcasters: The Art and Science of Prediction*. New York: Crown Publishers, 2015.

Theopedia.com. Entry on "Decisionism." http://www.theopedia.com/decisionism (accessed April 12, 2017).

Thomason, Richmond H. "Indeterminist Time and Truth-Value Gaps." *Theoria* 36 (1970): 264–281.

Thompson, Dennis F. "Election Time: Normative Implications of Temporal Properties of the Electoral Process in the United States." *American Political Science Review* 98, 1 (2004): 51–64.

Truffaut, François. *The Films in My Life* (translated by Leonard Mayhew). New York: Touchstone Books, 1985.

Turim, Maureen. *Flashbacks in Film: Memory and History*. New York: Routledge, 1989.

Voss, Daniela. "Deleuze's Third Synthesis of Time." *Deleuze Studies* 7, 2 (2013): 194–216.

Wiener, Norbert. *The Human Use of Human Beings: Cybernetics and Society*. New York: Doubleday, 1954.

Wikipedia article. "Go and Mathematics." https://en.wikipedia.org/wiki/Go_and_mathematics (accessed January 23 2017).

Wikipedia article. "Monte Carlo tree search." https://en.wikipedia.org/wiki/Monte_Carlo_tree_search (accessed April 12 2017).

Willatt, Edward. "The Genesis of Cognition: Deleuze as a Reader of Kant," in *Thinking Between Deleuze and Kant* (edited by Edward Willatt and Matt Lee). London: Continuum Books, 2009: 67–85.

Wright, Colin. "Event or Exception? Disentangling Badiou from Schmitt, or, Towards a Politics of the Void." *Theory and Event* 11, 2 (2008), journal online website, pp. N/A.

Wu, George. "Anxiety and Decision Making with Delayed Resolution of Anxiety." *Theory and Decision* 46, 2 (1999): 159–199.

Yakira, Elhanan. "Salomon Maimon and the Question of Predication," in *Salomon Maimon: Rational Dogmatist, Empirical Skeptic: Critical Assessments* (edited by Gideon Freudenthal). New York: Springer, 2003: 54–79.

Zhe, Li. "Lee Sedol's Strategy and AlphaGo's Weakness", and "Nobody Could Have Done a Better Job than Lee Sedol" (translated by Y. Tong, Chun Sun, and Michael Chen). massgoblog.wordpress.com (posted March 11 2016, accessed March 20 2016).

Zizek, Slavoj. *For They Know not What They Do: Enjoyment as a Political Factor*. London: Verso, 1991.

INDEX

Adorno, Theodor 43, 137–8
Agamben, Giorgio 127
Agency 202–3
AlphaGo 188–96
Anscombe, G. E. M. 9
Aristotle 31–2, 147, 159–60, 199–203
Artificial Intelligence 188–96
Austin, J. L. 130

Backward induction 160–1, 170, 173, 196
Badiou, Alain 5, 189, 240
Baho, James Scott 101 n.4
Bates, Jennifer 29 n.10
Beaufret, Jean 257
Belknap, Nuel and Michael Perloff 202–3
Benveniste, Émile 264
Bergson, Henri 50, 145, 206, 209
Berthoz, Alain 54
Borges, Jorge Luis 270
Braddon-Mitchell, David and Caroline West 210
Branching 150, 189–91, 196–223, 244, 247, 265–7, 269, 284–6, 290–2
Briggs, Rachael and Graeme Forbes 214–17
Broad, C. D. 204–6, 210, 265
Brunner, John 214–15
Butler, Judith 127

Cacciari, Massimo 240
Calculation 30, 108, 143–5, 155–96, 232–3
Cardozo, Benjamin 133–5
Chrétien, Jean-Louis 123
Christianity 105–25
Chu, John 214
Computation 241–3
Countermotives 74–79
Currie, Gregory 211–12
Cybernetics 210

Dark precursors 261–3
De Beauvoir, Simone 26
Decision point recognition 172
Decision Theory 41–2, 132, 155–88, 231–2, 248, 293
Decisionism 102–3, 117–20, 123–42, 145–6, 260
Deleuze, Gilles 209–10, 251–90
Derrida, Jacques 138–9, 243–9
Deterrence 171–2
Deutsch, David 217–23
Dice throw 266, 280–4
Diets 21–2, 32–47, 50–1
Disagreement with oneself 59–61, 79–82
Disjunction 58–61, 73–7, 99–100, 201–2, 206, 276–7
Divine foreknowledge 121–2
Duchamp, Marcel 287
Dummett, Michael 208–9

Effort 34–6
Egyptian Book of the Dead 114–15
Empty form of time 251–63
Equilibrium 178–82
Eternal return 259, 283
Eye in the Sky 237

Feedback loops 178–82
Fichte, Johann Gottlieb 75, 100, 267, 273–4, 277
Financial market predictions 182–7
Fixed (and unfixed) temporal position 66–70
Flashbacks and flashforwards 211–16
Forrester, Jay 178–82
Foton, Nick 151
Future contingents 199–202
Future generations 149–54
Future library 151
Fuzzy logic 173–5

Gambling 22
Game theory 155–87, 281
Gaps in the law 134
Gell-Mann, Murray 222
Gerrans, Philip 160
Gibson, William 292
Go game 188–96, 281, 287–8
Good Judgment Project 186–7
Growing Block model of the future 204–8

Habermas, Jürgen 139–42
Heap, Sean P. Hargreaves and Yanis Varoufakis 155–73
Hegel, G. W. F. 75, 78, 225–36, 240, 267, 274–7, 291
Heidegger, Martin 85–103, 107, 126 n.1, 133, 209, 213, 247, 272, 281
Henri, Michel 56
Henry, Claude 175–7

Hölderlin, Friedrich 257–8
Hope 112–15
Horwich, Paul 199, 207
Houle, Karen 237
Hume, David 147
Husserl 53–83, 89–90, 142, 154, 267, 279–80, 284

Improvisation 177–8
Incompossible worlds 284–90
Indecision 50–1, 65–6, 74–9, 83, 89, 149, 225, 236–41
Indifference 99
Infinitely repeated games 169–71
Iterated games 167

James, William 34–6, 289
Jarry, Alfred 131
Jaspers, Karl 26–7
Juxta-duration 71–2

Kahnemann, Daniel 156–7, 183
Kant, Immanuel 67, 112–14, 150–1, 226–8, 233, 256–7, 267, 271–5
Keynes, John Maynard 184–5, 187
Kierkegaard, Søren, 106–9, 115–24, 131, 249

Leap of faith, 107–9
Legal decisions 125, 133–5, 139–42
Leibniz, Gottfried Wilhelm von 198, 284, 287–9
Lewis, David 69, 198, 211, 217–18, 285–6, 289
Lewis, Jerry 283
Linz, Juan J. 147–8
Lodzinsky, Don, 120–1
Luhmann Niklas 142–6
Lyotard, Jean-François 43

Machiavelli, Niccolò 147
Many worlds interpretation of quantum physics 216–23

Marion, Jean-Luc 123–4
Masaki, Takemiya 287
Mbiti, John 44–5
McCall, Storrs 206–8
McDermott, Michael 159
McTaggart, John McTaggart Ellis 212
Meinong, Alexius 68 n.6
Meyer, Lukas H. 152
Molina, Luis de 121–2
Moments 117–19
Multiple-criteria decision analysis 174

Nahin, Paul 215–16
Newcomb problem 161–4
Nietzsche, Friedrich 29–30, 209–10, 283–4
Noema 23, 32, 55–62, 83, 267, 279–80, 284, 289
Nonlinear behavior 178–82

Oaths 24, 108
Object = X 256, 266–80, 284, 288
Ockham, William 79–80
Overlap 62–66

Pascal, Blaise 109–12
Pasolini, Pier Paolo 213
Performatives 130, 202
Peris, Daniel 182–3
Peterson, Martin 155–73
Plans 8, 146–7, 159–60, 177–8
Plotnitsky, Arkady 263
Position-taking 54–60, 72–78, 82
Possible worlds 197–8, 217–18, 267, 284–90
Potential 200–1, 291–2
Price, Huw 199
Prisoner's dilemma 164–73, 293
Promises 25, 29–30, 107–8, 123, 167
Prospect theory 156–7, 183

Quasi-time 66–70

Randall, Lisa 214
Rawls, John 153–4
Redmond, Michael 188–90, 193–4
Resoluteness 101–2
Reversibility 175–7, 184, 187
Risk 142–6, 157, 185
Robarts, John P. 147
Rommelfanger, H. 173–5
Rousseau, Jean-Jacques 147
Ruiz, Raul 281 n.40

Santiso, Javier 148–9
Sartre, Jean-Paul 21–51, 77, 133, 293
Saving for the future 154
Schelling, Thomas C. 166, 171–2
Schmitt, Carl 126–33, 240, 249
Schmitter, Philippe C. 148–9
Science fiction 150, 198, 211–16
Scotus, John Duns 79–80
Sedol, Lee 188, 193
Sequential decisions 159–87
Sexuate undecidability 237
Shrinking tree model of the future 204–8, 217
Simak, Charles 215
Smart, J. J. C. 215–16
Sokolowski, Robert 23, 121–3
STIT (See To It That) theory 202
Striving for a decision 58–60, 73–7, 81–2
Szendy, Peter 150–1
Supertemporality 72–4
Systems theory 178–82

Taleb, Nassim Nicholas 184
Temporal asymmetry 198–9
Temporal phase pluralism 210
Temporally disjunct inserts 211
Tense 197–215, 264–6
Term limits 146–9
Tetlock, Philip 186–7
Tewari analysis 287–8

Thomason, Richmond H. 202–3
Thompson, Dennis F. 149
Thrasymachos 136
Time preference 153–4
Time travel 198, 211–16, 220–1
Trembling hand hypothesis 171
Truth-value gaps 202–3
Turim, Maureen 211
Turing Decision Problem 241–3
Turning points 267–70

Uncertainty 110–12, 143–6, 157–9, 173–84
Undecidability 239–49

Unintended consequences 178–82
Until further notice 142
Utopias 8

Virtual ethics 150–3, 288–93

Weakness of will 31–2
Wells, H. D. 214
Wiener, Norbert 210
Will to believe 106, 110
Wu, George 177

Zhe, Li, 195
Zizek, Slavoj 1

www.ingramcontent.com/pod-product-compliance
Lightning Source LLC
Chambersburg PA
CBHW071759300426
44116CB00009B/1135